PHYSIOGRAPHIC UNITS - VALENCIA

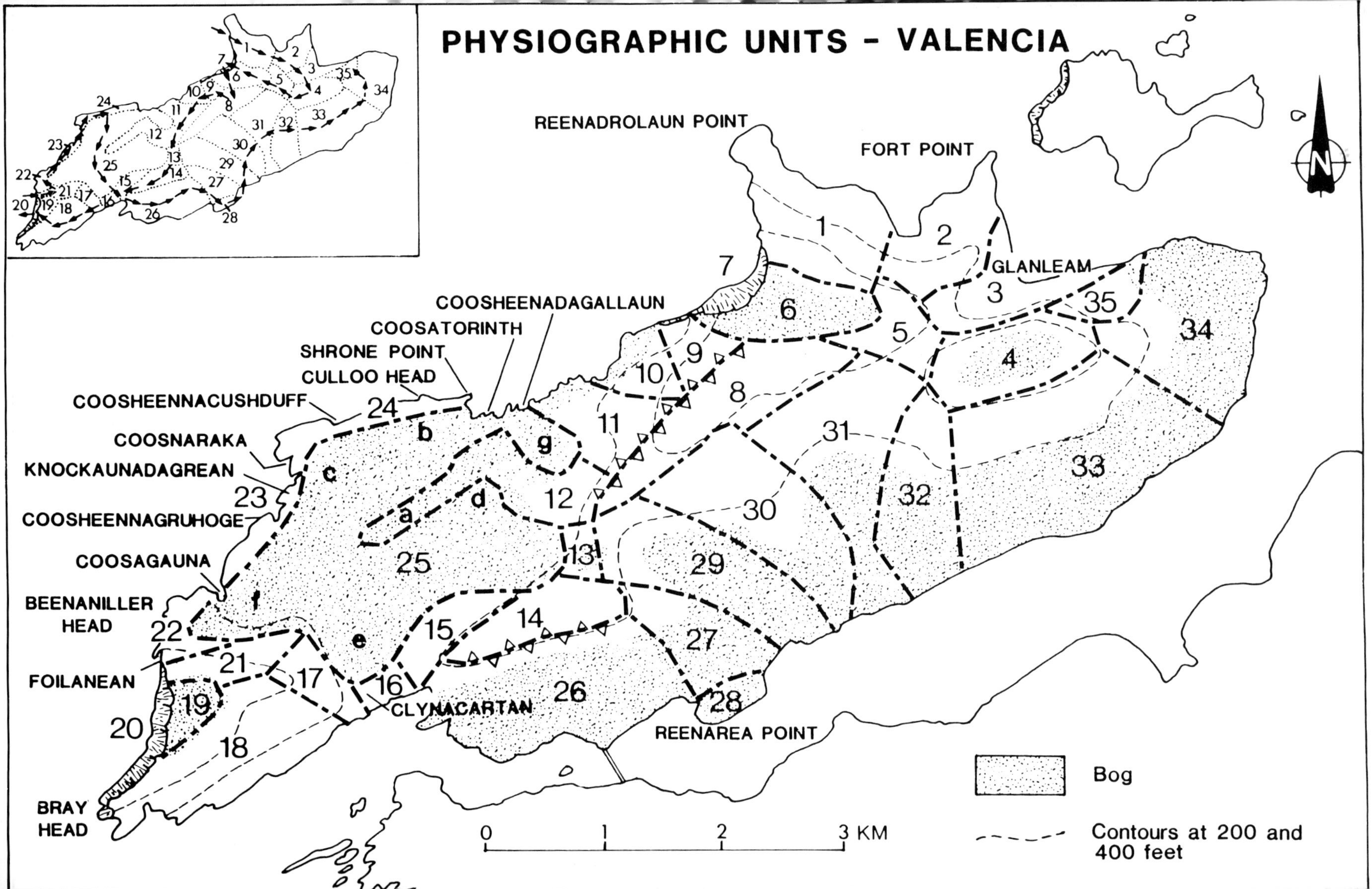

PHYSIOGRAPHIC UNITS

Numerical order

1	Dohilla Slope
2	Fort Point
3	Glanleam
4	Kilbeg Top
5	Gortgower Saddle
6	Geokaun Top
7	Fogher Mega-Cliffs
8	Kildreenagh Dip-Slope
9	Tinnies Scarp-Slope
10	Cool Scarp-Slope
11	Cool East Lobe
12	Cool West Lobe
13	Coarhmore Saddle
14	Coarha Top
15	Coarbabeg Lobe
16	Clynacartan Saddle
17	Bray East Slope
18	Bray South Dip-Slope
19	Bray Top
20	Bray Mega-Cliff
21	Bray North Scarp-Slope
22	West Rock Rim
23	Coastal Downland
24	North Rock Rim
25	Imlagh Basin
a	Main Escarpment
b	North Rock Rise
c	Brendan Rock-Knoll
d	Emlagh Bog Embayment
e	Clynacartan Bog Embayment
f	Bray West Bog Embayment
g	Cool Bog Embayment
26	Dysert Slope
27	Oonagranshee Lobe
28	Reenarea Rise
29	Donnybrook Valley
30	Tinnies Lobe
31	Chapeltown Valley
32	Ballyhearny Lobe
33	Laharan Dip-Slope
34	Knightstown Nose
35	Kilmore Shelf

Alphabetical order

Ballyhearny	32
Bray East Slope	17
Mega-Cliffs	20
North Scarp-Slope	21
South Dip-Slope	18
Top	19
West Bog Embayment	25f
Brendan Rock-Knoll	25c
Chapeltown Valley	31
Clynacartan Bog Embayment	25e
Saddle	16
Coastal Downland	23
Coarha Top	14
Coarhabeg Lobe	15
Coarhamore Saddle	13
Cool Bog Embayment	25g
Cool East Lobe	11
West Lobe	12
Scarp-Slope	10
Dohilla Slope	1
Donnybrook Valley	29
Dysert Dip-Slope	26
Emlagh Bog Embayment	25d
Fogher Mega-Cliffs	7
Fort Point	2
Geokaun Top	6
Glanleam Pocket	3
Gortgower Saddle	5
Imlagh Basin	25
Kilbeg Top	4
Kildreenagh Scarp-Slope	8
Kilmore Shelf	35
Knightstown Nose	34
Laharan Dip-Slope	33
Main Escarpment	25a
North Rock Rim	24
Rise	25b
Oonagranshee Lobe	27
Reenarea Rise	28
Tinnies Lobe	30
Scarp-Slope	9
West Rock Rim	22

Man & Environment in Valencia Island

Man & Environment in Valencia Island

Frank Mitchell

Royal Irish Academy
Dublin 1989

ROYAL IRISH ACADEMY
19 Dawson Street
Dublin 2

ISBN 0 901714 77 1 (cloth)
0 901714 78 X (paper)

British Library Cataloguing in Publication Data

Mitchell, Frank
Man and environment in Valencia Island
1. Ireland. Geographical features, history
I. Title
911′.415

Cover design by Ann Creavin
Cover illustration by Aislinn Adams
Typeset and printed by Printset & Design Limited

To the memory of Augustus Henry Lane Fox (*later* Lt-Gen. Pitt-Rivers),
Deputy Quarter-Master General, Cork, 1852-5:
a dedicated field-walker.

Contents

Preface

In my earlier books on the Irish landscape I ambitiously tried to give an outline development of the countryside of the whole island from its first appearance in the geological past to the present day.

Ireland measures three hundred miles by one hundred and fifty. Now I have altered my focus, and I look at an area that is tiny by comparison, only seven miles by two. I have worked for ten years in Valencia Island, walking its hillsides and its bogs, and I now try to give a picture of the landscape as it has changed and developed from the time its basic geological foundation was first blocked out millions of years ago to its present state and to what will happen to it in the Single European Market.

At first I tried to base my field-work on the 6″ Ordnance Survey maps, and on the 2½″ derivative from them, which has both contours and the National Grid. But for much of the island there was not sufficient detail to enable me to position myself with any degree of accuracy. I then arranged for the Survey to provide me with overlapping aerial photographs on the scale 1/4000. Stereoscopic examination of the photographs revealed many new features, while the photographs — where every peat-cutting could be identified — were invaluable in the field. But I was unable to transfer grid references to the photographs.

My account of the island varies in level, being perhaps over-detailed here and too sketchy there, but that is inevitable in the present state of knowledge. An explanation of such technical terms as are unavoidable may be found in the glossary. I have restricted the bibliography to references that have an immediate bearing on the island and its history.

Mesolithic hunter-fishers arrived on Valencia six thousand years ago, and I follow the human history of the island through the coming of Neolithic farmers, the Bronze Age people who laid out field systems which can still be traced, the Early Christian folk who established monasteries and farms throughout the island, the land changes of the sixteenth and seventeenth centuries, and the potato-covered landscape of pre-Famine days.

Much of the record comes from the island's bogs, which expanded steadily down the millennia, swallowing up man's walls and buildings. Peat-cutting brings them to light again.

The history of man and environment in Valencia is a fascinating one, and nature has recorded it in remarkable detail. Although this book is overlong, it is still no more than a preliminary survey. I trust that others will follow, and reveal more of the remarkable detail that is still stored up on the island.

Acknowledgements

I received generous help from a great many people, and I trust that my list of acknowledgements is complete. For very special support I am indebted to Pat and Des Lavelle, who helped in a thousand ways; to Pete Coxon for his pollen studies; to John Sheehan and the South-West Kerry Archaeological Survey who helped in the field and on the drawing-board; to Michael Conry for discussions on soils; to John McCarthy who examined some souterrains; to Daphne Pochin Mould, who not only put her book on the island and her photographs at my disposal but also took me flying with her; to John Danaher of the Ordnance Survey who took the aerial photographs on which my field-work depended; to Adrian Phillips and MAPTEC for access to photogrammetric equipment; to John (Dorney) O'Connell and Nóra Ní Shúilliobháin for much local information; to Elizabeth FitzPatrick for many useful suggestions; and to Barbara Young and Emer Condit for their skilful editing.

I also wish to thank the many other colleagues and friends who from time to time helped me in the field, in the laboratory, at the drawing-board, by the loan of photographs, or by discussion. I particularly call to mind Fred Aalen, Cathy Barnosky, Dick Bryant, Seamas Caulfield, Catherine Coxon, Mary Davies, Kim Donner, George Eogan, Tom Fanning, Patricia Johnson, Diana Large, John Kelly, Ann Lynch, Phelim Manning, Tessa O'Connor, Eugene O'Sullivan, and Maura Scannell.

The work was supported by grants from the Stacpoole Archaeological Fund of Trinity College, Dublin, the University of Dublin Fund in the United States, the AnCO (now FÁS) / South-West Kerry Development Organisation, the Geological Survey of Finland, and a generous grant towards the cost of publication provided by Lennox Chemicals.

Illustrations

Illustrations are referred to by the number of the page on which they appear. Cross-references are shown by the relevant page number in italics on the page margin. The source of the illustrations is shown as follows:

Anon.	Artist unknown
DEL	Des Lavelle
DL	Diana Large
DPM	Daphne Pochin Mould
EEE	E. Estyn Evans
EOS	Eugene O'Sullivan
FM	Frank Mitchell
GE	George Eogan
OS	Ordnance Survey
PMJ	Patricia Johnson
PM	Phelim Manning
RH	Robin Holder

Permission to reproduce is gratefully acknowledged.

The type of illustration is shown as follows:

LD	Line-drawing
BW	Black and white photograph
COL	Colour photograph

1
Introduction

In south-west Ireland folded mountain ridges slope down gently, owing to tectonic subsidence, till they disappear below the ocean. The Iveragh Peninsula, of which Valencia Island is a detached fragment, has many such rock ridges, separated by intervening valleys, all running north-east/south-west.

In Iveragh the rock everywhere is Old Red Sandstone, which coarsens in places into conglomerate, and fines in others into slate. Earth pressures created the ridges and valleys and impressed a cleavage on the rocks, which in consequence often split at a high angle to the original sedimentary layers. This is well seen in the great slate-quarry at Dohilla at the northern end of Valencia. The rocks are highly siliceous, and soils derived from them are deficient in nutrient elements.

Valencia is a narrow island, about 11km long and 3km wide, aligned east-north-east/west-south-west. It is essentially a detached piece of the mainland rather than an island in its own right. The elongate Portmagee *2*
Channel on the south-east is nowhere more than 4m deep. At its northern end, between Knightstown and the mainland, a 10m-deep trench is cut into its floor. The Channel then runs north-west between Valencia and Beginish, creating the deep water of Valencia Harbour in the south-west corner of Doulus Bay. The rest of the bay is shallow, and its numerous islands suggest recent submergence. West of Portmagee, at the western end of the Channel, the water suddenly deepens to 20m, and the line of deepening is essentially in line with the eastern slope of Bray Head, probably itself a tectonically uplifted block or horst; I call it the Bray Horst. With the opening of a bridge at Portmagee in 1971, Valencia has lost its island status.

The climate today can fairly be described as very equable. The annual value of mean daily air temperature is 10.8°C; in monthly values August (15.4°C) is the warmest and February (6.8°C) the coldest. Frost occurs on 42 days of the year, and snow lies on the ground for one day only. Grass growth continues the whole year round. The sky is often cloudy, and the annual average of daily duration of bright sunshine is only 3.72 hours. May has the most sunshine and December the least. Annual rainfall amounts to 1400mm, almost twice that of the Dublin area. It falls on 190 days, with December (168mm) being the wettest month and April (74mm) the driest.

Wind is the chief feature of the local climate: it reaches gale force on 123 days in the year, it can reach a speed of 90 knots, and it usually comes from the south-west quarter. With most of the air moving in off

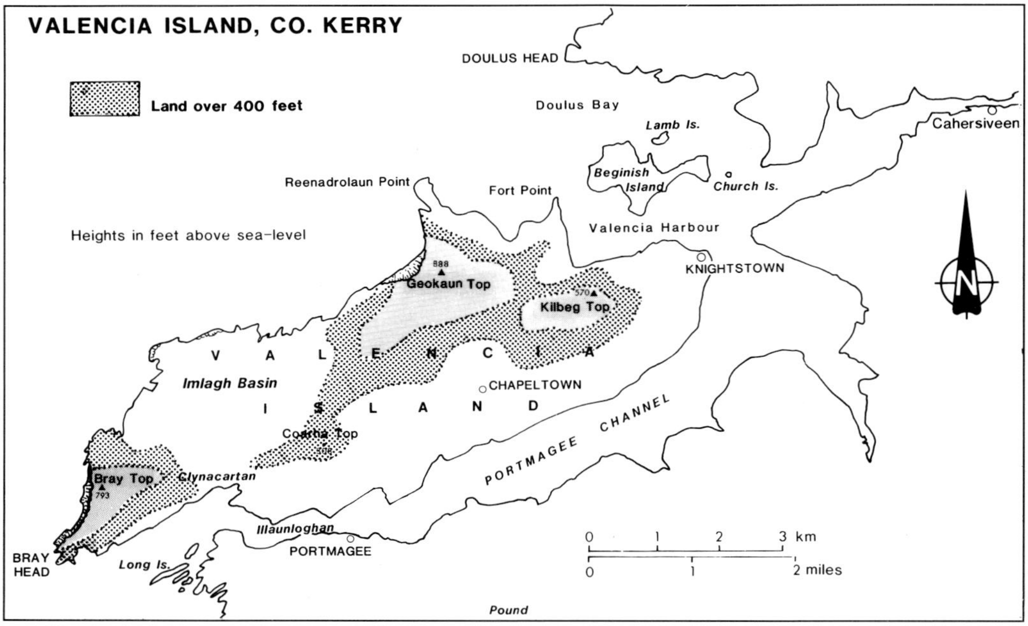

2 *General map of Valencia Island.*

the Atlantic Ocean, humidity is high and evaporation is low. These conditions make crop-drying very difficult and favour the spread of potato blight.

Such is the present picture. There are some indications that climate has not been constant throughout the six thousand years in which man has been present in Valencia, but the variations will not have radically altered its basic equable and windy qualities. For a time it may have been slightly warmer and drying conditions may have been better but, as far as farming is concerned, grass will always have been the most reliable enterprise. Today it is the sole enterprise, even further encouraged not only by high EC milk prices but also by the advent of silage-making, which has largely eliminated the risks of the hay harvest. Corn and potatoes, with their harvest hazards and high labour costs, have disappeared.

Physiographic features

A most important feature of the island is that there is no flat land. There
are slopes of every gradient, from steep rock cliffs to solifluction gradients
of 2° or less, but there is no flat land.

The northern third of the island can be regarded as an asymmetric,
tilted crutch. The concave head of the crutch runs from the highest point,
Geokaun (270m), eastwards to Kilbeg Mountain (175m). The tilted shaft 2
is attached not to the centre of the crutch-head but to its left-hand end,
springing from Geokaun, and running in an undulating ridge for 3km,
dropping in level to 90m as it does so. Here, a little to the south of the
end of the main ridge, the ground rises into an oval knoll, with top at
100m, the Coarha Top. The place-name Coarha indicates a look-out 2
point; from the Top the sea is visible to north and south, with Bray Head
to the west and the high ground to the east. From the Top the dislocated
ridge continues south-westwards for a further 1.5km, reaching the cliffs
on the south coast of the island at Clynacartan at an altitude of 40m. 2

The ridge forms, as it were, a spine to the island, and I call it the
Spine-ridge. When setting up the units into which I divided the
topography of the island, I regarded the ridge as a dividing line only,
without any lateral extent, although the ground it traversed sometimes
rose into minor domes, sometimes sloped away gently, sometimes steeply,
on either side. The south-eastern face of the ridge is essentially a dip-
slope, and its topography is relatively smooth, with some good soil. The
north-western face is a scarp-slope and shows much broken rock; there
has been little cultivation here. The cleavage in the rock is at a steep
angle to the dip, and this gives rise to rock overhangs.

South-west of Clynacartan the land rises abruptly into the Bray Horst,
a rectangular block (2km x 1km) with three straight edges and a cliff- 2
face, which reaches a height of 240m. The name Bray (*Brí*) means a
height.

North of the western end of the Spine-ridge and of the Bray block
there is an oval area of peat, about 4km long and 2km broad. At first
sight it appears to occupy a basin, and in earlier papers I have called
the area the Imlagh Basin. On its northern, coastal flank it is bounded 2
by a rim mostly of rock, except for one stretch of glacial gravel, nowhere
falling below 20m in height. Although the Irish word *Imleac* is commonly
used for a marsh or a bog, it can have a stricter meaning as 'edge of
land near cliffs' or 'fringe of bogland'. Perhaps in this area it originally
referred to the rim lying between bog and sea. I use the name *Imlagh*
(i.e. marsh, bog) for the whole basin, as this name is on the six-inch
OS map. At another point in the basin a site with a cross was named
Emlagh by earlier workers, and I use Emlagh for that particular site.

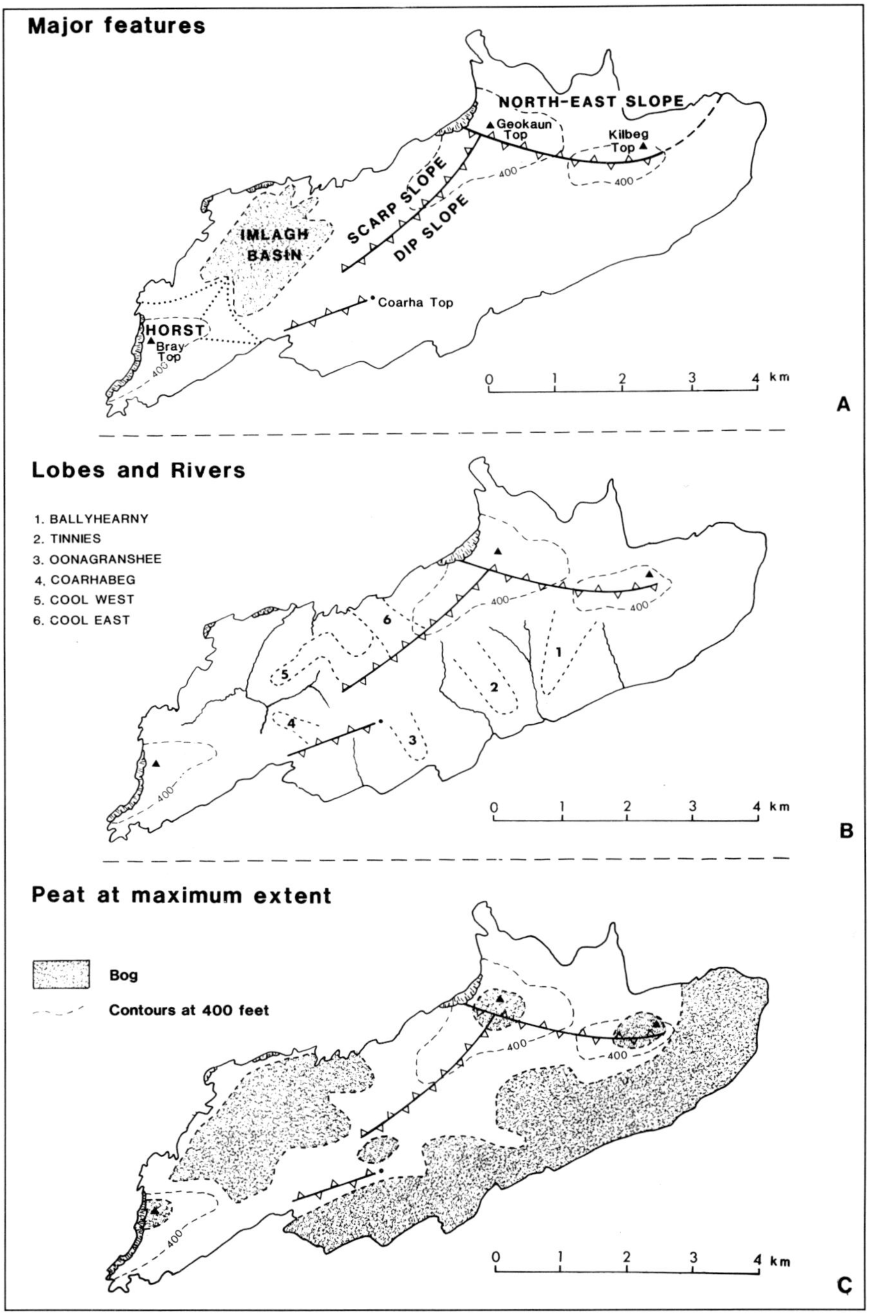

4 *Maps showing (A) major features; (B) lobes and rivers; (C) peat at maximum extent.*

In fact the 'basin' is a shallow arcuate valley in rock, as would be seen if the peat it now holds were completely removed.

The part of the island south-east of the Spine-ridge can be crudely regarded as a dip-slope which drops gently down until it disappears below the water of the Portmagee Channel.

These features are the major topographical units of the island, but I have attempted to recognise more minor units whose layout probably influenced man's choice when he was seeking sites for settlement on the island.

I am convinced that the rock topography underlying these units, both large and small, has a very long history, going back over many millions of years. I believe that the geomorphology, not only of the island itself but also of the country around Cahersiveen, Ballinskelligs and Waterville, is the product of profound tropical weathering that had its roots in the Tertiary period, perhaps about forty million years ago. Severe chemical weathering will have produced a deep mantle of residual soil resting on a relatively subdued rock surface.

After the rock topography had been shaped by chemical weathering, it was subjected to severe alternations of climate during the Quaternary. Its soil mantle was quickly stripped away, and the surface of the underlying unweathered rock was revealed as an etch-plain; this surface is the basis of the modern topography.

I also think that there was late tectonic movement, which gave rise
to regional uplift in south-west Kerry. Bray Head has a horst-like *35*
appearance, and the high coastal cliffs are very unstable.

Ice-masses probably formed over the mountains of central Kerry on more than one occasion, but the latest ice-advance, perhaps about 18,000 years ago, halted near Waterville and did not reach Valencia, though the island had been previously overrun by ice. On the north side of the island two sets of glacial striae have been recorded. One set runs north/south; erratics from Mayo can be found in the beaches north of Castlegregory in the Dingle Peninsula, and there is a large erratic of Galway granite on the beach at Ballinskelligs to the south; at some stage ice must have moved from north to south down the west coast of Ireland. A second set runs roughly east/west; around the estuary of the Shannon, striae and other indications of ice movement suggest that at one stage ice moved westwards offshore.

The ice did deposit glacial debris on the island, though much of this has been grossly disturbed by frost action during later cold periods. But along a short stretch of the north coast thick deposits of coarse gravel are relatively undisturbed. This area — very important to prehistoric
farmers — will be discussed later under the title *Coastal Downland.* *71*
Elsewhere freeze-thaw action caused most of the unconsolidated glacial debris to move downhill, where it accumulated at a low angle of rest; bare rock was exposed on the tops and upper slopes. If there were hollows anywhere in the rock in which lakes might later have formed, these were filled in. Even after the moved debris had reached the angle of rest, its

upper layers were churned around by later freeze-thaw cycles and splendid involutions were formed at many places.

Given the basic structure of the island, I have subdivided the physiography into small-scale units. I regard these as having been shaped by tectonism, by prolonged sub-aerial erosion, by alternating heat and cold operating on a much shorter time-scale, and by post-glacial soil development.

The units are shown on a map, which appears as endpapers. Attached to the map are lists showing the names I have given to the units, both in the sequence in which they are described and in alphabetical order.

My units are perhaps over-elaborate, but they may be grouped as follows.

Mega-cliff. André Guilcher has discussed the cliffs of western Ireland and identifies as 'mega-cliffs' those which are more than 150m high; he
54 especially noted those on Valencia, at Bray Head (230m), and at Fogher (240m) on the north side of the island. For their growth he pictured that the cutting by waves of a basal notch would bring about progressive cliff collapse. But steep slopes continue seawards below the cliffs, and in my opinion the instability of the cliffs is due to recent tectonic uplift.

33 *High Top.* High Tops are dome-like summits, covered by blanket-bog. There are three such features, Geokaun, Kilbeg and Bray.

35 *Low Top.* Low Tops are of lesser elevation, not exceeding 35m. Coarha Top at 33m shows three rock ridges separated by gullies. Reenarea Rise at 15m is a low peat-covered dome.

35 *Saddle.* Saddles lie between tops. The Gortgower Saddle links Geokaun and Kilbeg Tops; the Clynacartan Saddle links the Spine-ridge to Bray; the Coarhamore Saddle spans the break in the Spine-ridge.

4 *Lobe.* Lobes are down-stretching ridges of higher ground which protrude from the Spine-ridge, or from Tops. They are usually free of glacial debris, and have well-drained soils, attractive to farmers. Their upper slopes remained free of peat.

4 *Valley.* Because the topography was formed by chemical weathering and not by fluvial erosion well-defined valleys are few, and much of the surface water is carried off by small rills.

Single-stream valley. The Donnybrook Valley and a very small valley (not reckoned as a separate unit) on the Dysert Slope carry small streams.

Double-stream valley. At Chapeltown two streams which have followed the sides of a 'V' unite and flow seawards.

Arcuate-stream valley. The bog that fills the Imlagh 'Basin' conceals an arcuate valley. A projection from a north-west-running lobe (the Cool West Lobe) from the Spine-ridge first falls in level and then rises into an elongate rock ridge running north-east/south-west. The drainage stream, at first flowing north, was compelled to turn south-west, and then to make a hairpin bend round the western end of the ridge, and it still follows this arcuate course even though the rock ridge later became completely buried by peat growth.

Pocket. At Glanleam there is a deep, steep-sided small hollow open to the sea on the east, the Glanleam Pocket. Further east, across Valencia Harbour, the Cahersiveen River has a remarkably straight course, overhung on the north by the steep southern slope of Knocknadobar, and this topography is thought to be due to a tectonic fault. The fault continued south-west to Glanleam, and the steep southern wall of the Pocket (dipping at 60°) is a direct consequence of the fault movement. Sub-tropical plants grow freely in the shelter of the Pocket.

Shelf. At Kilmore there is a small even shelf, sloping gently to the east-north-east, cut into the east side of Kilbeg Top and cut off to the north-west by the Glanleam Pocket. Its geomorphological development is not clear.

Slope

Dip-slope. On the south-east of the Spine-ridge, on its continuation *44*
from the Coarha Top, and on the southern slope of Bray Head there are dip-slopes. On the upper part of the slopes, where the gradient is 15-10°, soils are well drained and — suitably manured — can be fertile. On the lower slopes, where the gradient is 9-2°, and where the solifluced rock debris has become consolidated by movement, drainage is poor. In addition, water that has drained down the upper slope below the surface rises upwards as the slope decreases and makes the lower slopes

TABLE 1
Classes of physiographic units

Unit class	Number of units	Reference number on map
Mega-cliff	2	7, 20
High Top	3	4, 6, 19
Low Top	2	14, 28
Saddle	3	5, 13, 16
Lobe	6	11, 12, 15, 27, 30, 32
Valley	3	25, 29, 31
Pocket	1	3
Shelf	1	35
Slope		
Dip-slope	5	8, 18, 24, 26, 33
Scarp-slope	3	9, 10, 21
Other slopes	6	1, 2, 17, 22, 23, 34

still wetter. Extensive bog formerly covered the lower slopes, but most of this has now been cut away and the poor soil has re-emerged.

54 *Scarp-slope.* On the north-west of the Spine-ridge, north of Coarha Top, and on the northern slope of Bray Head there are scarp-slopes. Gradients tend to be steep and rock outcrops are numerous, giving rocky broken ground with a very poor soil cover. There has been no cultivation on these slopes, which carry a thin layer of peat.

Other slopes. Some slopes, such as Dohilla, Fort Point, Bray East Slope and the Coastal Downland, do not fall into these categories.

Vegetation since the Ice Age

Having outlined the topographical features to be met with on the island, I turn to the way in which plants have clothed them since the Ice Age.

The record of vegetational history has not been built up in a systematic way, with a master-site to which all other sites are referred. As a result of peat-cutting and drainage no master-site survives, and the story has had to be built up by assembling data from a number of sites, randomly discovered, which had archaeological features but only a partial vegetational record.

Also during the years I was working on the island environmental studies were not standing still. It was becoming apparent that at least as far back as the Bronze Age there were very extensive areas of farmland, and that much light could be thrown on the farming practices of the day by detailed pollen counts at close sampling intervals. To these developing studies I have not made any significant contribution.

Pollen counts were made by Pete Coxon at five sites in different parts of the island, and a number of radiocarbon datings (Table 2) were obtained, both at these sites and some others. All the ^{14}C dates cited are 'raw' dates which have not been calibrated. Mick O'Connell and Peter Foss also made some pollen counts.

Pete Coxon's main pollen diagrams, two shorter diagrams by Mick O'Connell and Peter Foss, an older diagram by Jessen from Emlaghlea on the mainland nearby, 10km to the south-east near the Cahersiveen/Waterville road, and a recent diagram by Ann Lynch from Cashelkeelty in south-west Cork, 40km to the south-east, are now presented with explanatory captions.

Reading from left to right the diagrams show first the depth and stratigraphy of the section, then the relative percentage values for trees and bushes set out in the order in which they probably entered the country after the Ice Age, then grasses, sedges and heathers associated with bog development. On the right is a disturbance curve in which those weeds and other plants generally taken to be associated with human activity, such as plantains, docks, dandelions, thistles and nettles, are united. Radiocarbon dates are on the extreme right. Descriptions in the captions read upward from the base of the section. Owing to re-drawing and reduction, the curves in the diagrams should be regarded as form-lines to show trends, and not as accurate contour-lines.

Valencia today has no lakes, and in my wanderings over the island I did not see anywhere fossil lake-muds exposed either at the base of peat-cuttings or on the floor of the drainage stream in the Imlagh Basin. The natural course of the stream has been very much altered by drainage cuts, which in many cases go down to glacial debris or bedrock. As I

TABLE 2
Radiocarbon dates

With the exception of the first date (H 2159), which was done in Helsinki, all the other datings were carried out by Teledyne-Isotopes. From left to right, the table gives:

Code — The serial number of the sample.
Site — The general location of the sample site, abbreviated as
BT = Bray Top,
I = Imlagh Basin and its rim,
LS = Laharan Slope,
RR = Reenarea Rise.
To assist the location of sites in the Imlagh area, the unit number follows the letter I.
Mat. — The material of which the sample was composed, abbreviated as:
C = Charcoal,
M = Mor,
P = Peat,
S = Shell,
W = Wood.
Age — The ages are given in uncalibrated radiocarbon years.
Description — A very brief statement: detail will be found in the relevant unit.
Arch. horizons — The date indicates the times at which successive archaeological phases are thought to have opened in Ireland.

Code	Site	Mat.	Age	±	Description	Arch. horizons
H 2159	I12	S	930	80	Shells in souterrain	1550 EC
14209	I25e	P	1700	80	At level of earth track	
13911	I25c	P	1720	80	At level of stone track	
14081	I25	P	2020	80	Below earth track	
15388	I23	P	2050	80	Above LBA wall	2450 IA
15098	I23	P	2680	90	At level of LBA fence	2700 LBA (Dowris)
15387	I23	M	2900	90	Below LBA wall	
15545	I25	P	2920	80	At level of stone track	
15470	RR	P	2950	80	Below EC cooking-place	
15385	LS	C	2970	90	In disturbed cooking-place	3150 BA (Bishopsland)
13751	I25f	C	3160	90	In small hearth	
14083	I25a	C	3250	100	In stone-built cooking-place	
13912	I25d	C	3400	90	In small hearth	3750 BA (Food Vessels, Urns)
15100	RR	C	3800	100	In large hearth	
15099	RR	W	3850	100	Immediately below large hearth	
14208	I25d	P	3930	90	At level of stone track	3950 BA (Beaker)
13752	I25d	W	4600	100	Oak stump in fen-peat	
14206	I25d	W	4760	100	Willow twigs at wall base	
15607	BT	M	4790	120	Mor at base of peat	
15469	RR	P	4950	110	At disturbance horizon	
14207	I25d	P	5120	100	Base of fen-peat	
15224	RR	P	5630	110	Base of fen-peat	5700 Neolithic
14082	I25e	P	6560	120	At level of stone platform	
13941	I25e	W	8910	150	Oak baulk in platform	9000 Mesolithic

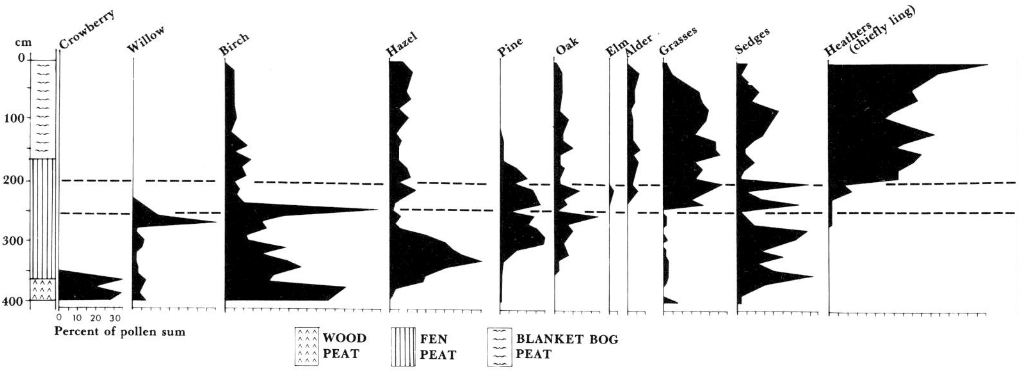

Emlaghlea Bog, near Waterville, Co. Kerry (after Jessen 1949, pl. XV, 48, detail added.). *11*

The section here is 400cm deep. Crowberry was still common when birch debris started to build up wood-peat about 10,000 years ago. The area became wetter and fen-peat with sedges and Phragmites *started to form at 365cm; the peat has a varying content of wood. Hazel then arrived and dominated the countryside till it had to give way on the arrival of the pine and the oak. At 240cm the dramatic fall in pine and oak, with a rise in birch immediately followed by a rise in grass, indicates interference with the woodlands by Neolithic man, before the arrival of the alder. A little higher, the arrival of heather pollen shows that blanket-bog was beginning to be established in the neighbourhood, and at 165cm blanket-bog vegetation replaced the fen; pine disappears at this level. At this level also a peat-cutter discovered a bronze axe with a stop-ridge, for which 3000 BP would be a plausible date. After this, blanket-bog flourished in the area and tree growth was very limited. The values for hazel pollen may be incorrect, as its pollen is easily confused with that of bog myrtle* (Myrica gale), *which grows freely on blanket bog.*

noted earlier, debris moved downslope by frost action will have filled up any pre-existing basins.

But basins containing early lake-muds do exist in south-west Kerry, and these record the return of life after the Ice Age had ended. Two of these have been closely examined in recent years. One is 20km south-east of Valencia, at Baslikane Bridge immediately south-east of Waterville, and the second 30km to the north at Ballinloghig in the Dingle Peninsula. The records from these sites show that the return of vegetation to the area started about 13,000 years ago and followed much the same pattern as that in the rest of Ireland. At first the plant life probably consisted of very sparse scattered herbs in a landscape in which bare soils predominated. Climate improved rapidly, soils developed, and birches, willows and a rich growth of herbaceous plants reclothed the landscape. Cold then made a short but severe return, and all but the hardiest forms of life disappeared.

About 10,000 years ago temperature started to rise rapidly, and soon reached a level as warm as, if not slightly warmer than, that of today. Plants and animals crowded back into Kerry. Hardy heath plants were the first to expand, and prominent among these was the crowberry (*Empetrum nigrum*). Willow and birch were the first trees to immigrate,

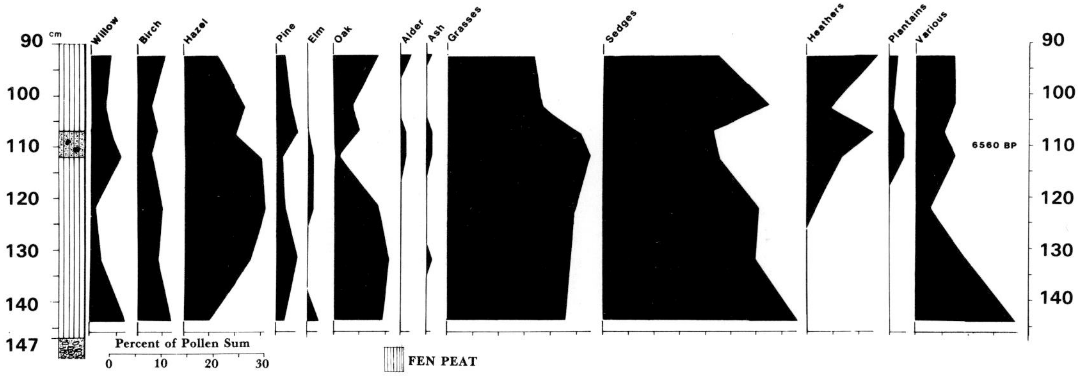

12 *Clynacartan Bog Embayment (to be published later in more detail).*
Though some peat had been removed, about 150cm remained. The top 60cm was humified blanket-bog peat; there was then 10cm of coarse peat rich in rhizomes of Phragmites, *resting on 80cm of* Phragmites-*peat, which became muddy at its base. A few metres south of the profile slabs of stone (interpreted as a platform) protruded from the peat face at a depth of 110cm. A thin sheet of sand and small stones (shown in the diagram) extended out from the platform for a short distance. The peat in which the debris lay had a* ^{14}C *date of 6560 BP. Pollen samples were counted between 92cm and 144cm.*

That the peat accumulated in a fen is shown by the high content of grasses and sedges. Impedance of drainage by peat formation further down the Basin probably caused ponding of water here about 7000 BP, and muddy fen-peat started to build up in a reed-swamp. Such conditions are suggested by occasional pollen of meadow-sweet (Filipendula ulmaria), *bur-reed* (Sparganium *sp.) and alder-buckthorn* (Frangula alnus). *The falling value in the curve for various pollens perhaps indicates the impoverishment in variety of the plants present as the reed-swamp gave way to full fen.*

*High forest of oak, hazel and pine occupied the dry ground around the fen, and ash, elm, yew and ivy were also present. At the level of the sand layer some interference with the woodlands is clearly seen. Elsewhere in the British Isles substantial evidence is accumulating that Mesolithic man did bring about woodland change, but what his activities were is not clear. There was interference here; oak and hazel fall back, plantains (including occasional ribwort plantain—*P. lanceolata*) and ash appear. Oak then recovers.*

The arrival of heathers (chiefly ling) suggests the beginning of blanket-bog formation in the vicinity. At the top of the diagram the establishment of alder is suggested.

and they quickly overshadowed the heaths. This development is recorded at Ballinloghig, where it was dated to about 10,000 years ago. In the
11 Emlaghlea diagram we can see the disappearance of the crowberry.

Hazel then arrived, quickly pushing the willow and birch aside, and 9000 years ago it dominated the countryside, producing enormous quantities of pollen. But its acme was short as pine, oak and elm were already moving into Ireland, and it was quickly reduced to the status of an underbush in tall woodland. Today it no longer grows on the island.

In Valencia a baulk of oak, later utilised in a Mesolithic structure, had a ^{14}C age of 8900 years, showing that the tree was in the island at that time. The baulk may have come from a tree which had grown on mineral soil near the margin of the bog, as a pollen count from the base

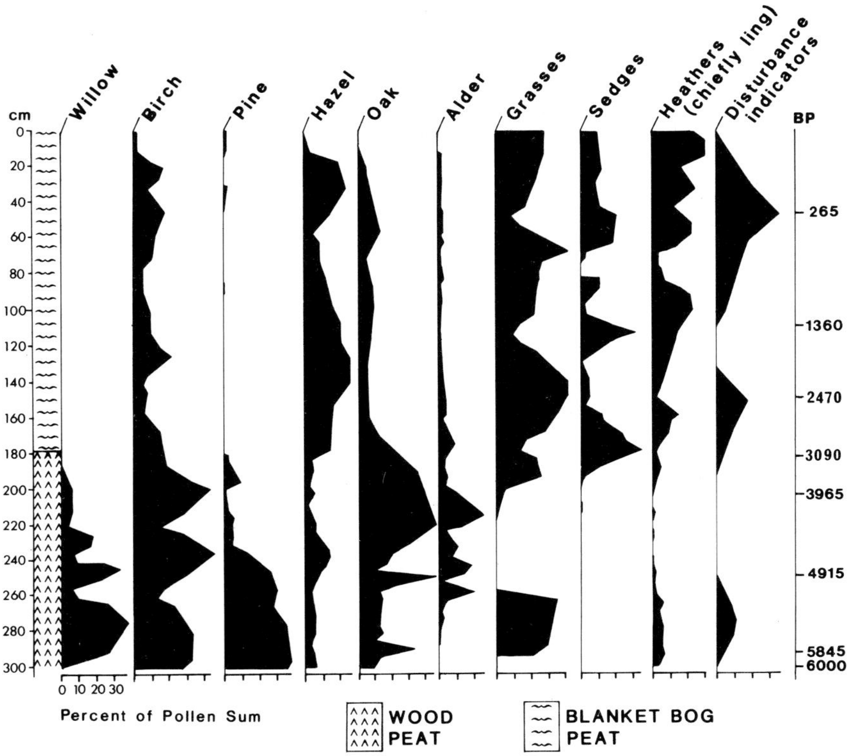

Cashelkeelty, Co. Cork (freely adapted from Lynch 1981). 13
Cashelkeelty lies in south-west Cork, 40km south-east of Valencia. The bog is developed on a rock-ledge at 110m on the peat-covered mountain slopes that run down to the south side of the Kenmare estuary. The particular site (Cashelkeelty I) is a small peat-filled hollow in sandstone.

The section here is 300cm deep.

The basal 120cm were a wood-peat, which started to form about 6000 BP when high forest dominated the site. Man started to create a clearance in the woods 5850 years ago, and grasses flourished. Heather and disturbance pollens (including traces of cereal pollen) were also present. About 4900 years ago disturbance ceased, and oak and alder grew freely. This disturbance must have been brought about by Neolithic man. Pine disappeared about 3100 years ago. A second phase of disturbance took place between about 3100 and 2500 years ago, and this lies where the Earlier was giving way to the Late Bronze Age. Oak and alder fell away, and grasses and disturbance pollens were frequent.

Blanket-bog began to develop at about the same time as the disturbance began, and human activity, forest reduction and the beginning of blanket-bog formation may all be interconnected. More intensive agriculture began about 1350 years ago, probably indicating the introduction of new farming practices from Europe as Christianity spread. This disturbance reached a peak about 250 years ago, when Sir William Petty was active in the Kenmare area.

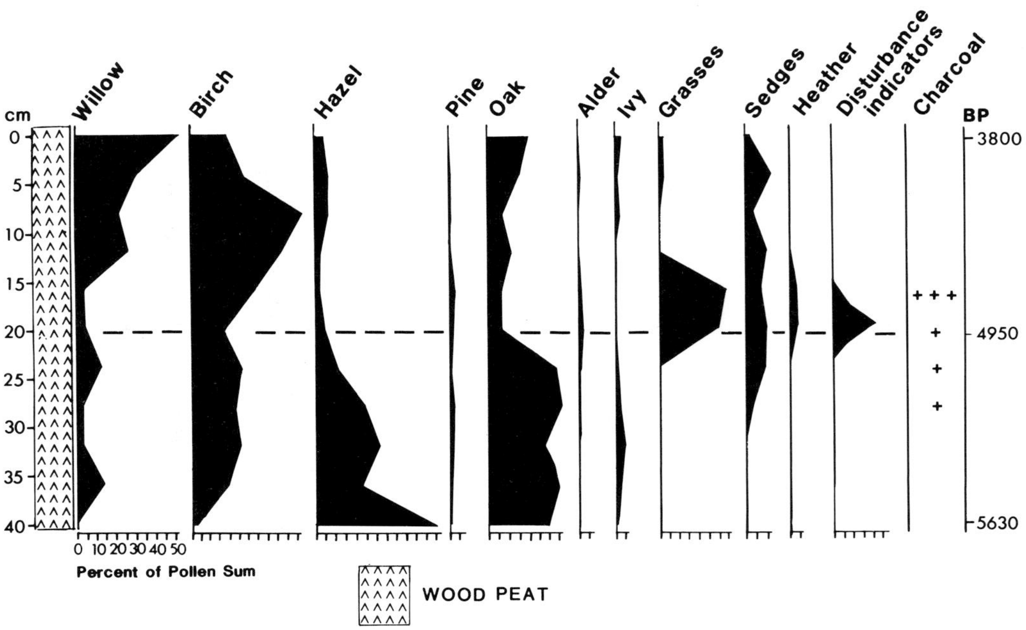

14 *Reenarea Rise (to be published later in more detail).*
The section here is 40cm deep.

The peat is a wood-peat, lying between a fulacht fiadh *dated by* ^{14}C *to 3800 BP and the base of the peat dated to 5630 BP. The lower peat shows high forest of birch, hazel and oak, with ivy and some pine. The appearance of sedge pollen at 30cm may indicate that blanket-bog was starting to form in the vicinity.*

Then at 20cm abrupt Neolithic woodland clearance—dated by ^{14}C *to 4950 BP—opens up a space in the forest, grasses and disturbance pollens leap up, and heather is present. Charcoal is present in the samples. Agriculture then ceases, and willow and birch quickly colonise the abandoned grassland. Some forest with hazel, pine, oak and ivy continues.*

of the peat below the structure suggested an age of the order of 7000 years. Elm did not like the poor soils of south-west Ireland, and it can never have been an important tree there. High forest, with pine and oak important, hazel, birch and willow common, and yew, holly and ivy also present, will have covered much of the Imlagh Basin and probably most of the lower slopes of the island as well. Such woodland
12 is documented for us in the pollen diagrams from the Clynacartan Bog
14 Embayment and Reenarea Rise, and also in the lowest sample from the
15 Brendan Rock-knoll site in the Imlagh Basin where oak, hazel and ivy are dominant with pine and yew nearby.

With the exception of the alder, all the common native trees of Ireland were now present in the area. It is thought that alder had become established over most of the country by about 7000 years ago, but it
11 did not arrive at Ballinloghig until 5250 years ago, and seems to have
13 been even more belated at Emlaghlea and in Valencia. At Cashelkeelty it became established about 5000 years ago. Oak was in the area by 9000

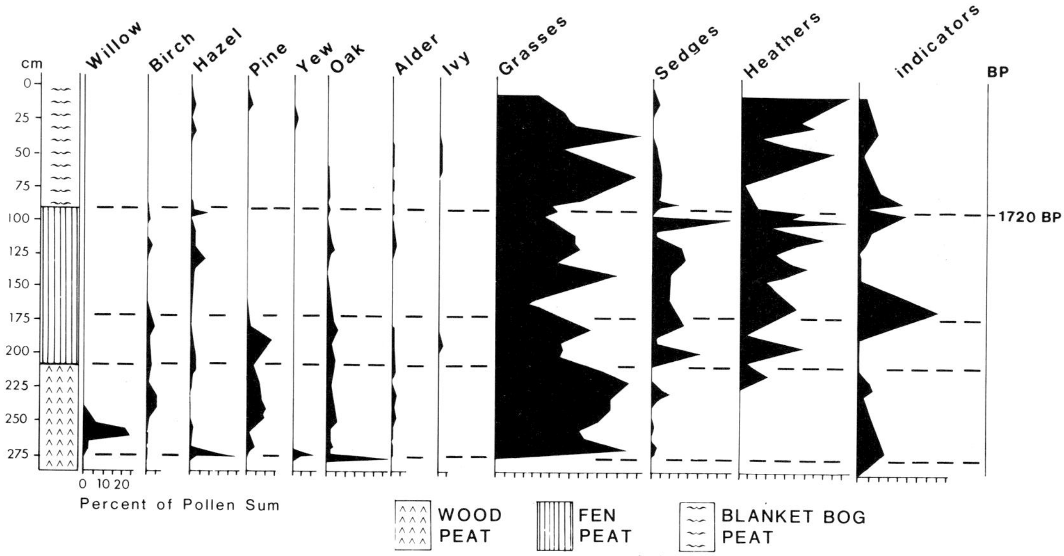

Brendan Rock-knoll (after Coxon 1985, 43). *15*
The section here is 280cm deep.

The lowest sample at 275cm, from peat rich in wood, shows high values for hazel, yew and oak, indicating high forest in the vicinity. These values immediately drop, grasses surge up and disturbance pollens appear, indicating Neolithic forest clearance. Disturbance pollens then fall away, grass pollen remains dominant, and tree values are low, never to become high again. Here we probably see the consequence of pre-3000 BP tree clearance on the Coastal Downland immediately to the west, and the establishment of grassland there. At about 215cm the basal wood-rich peat gives way to fen-peat, and the appearance of heather and sedge pollen in the counts shows that blanket-bog was beginning to establish itself in the vicinity. At 170cm trees disappear, disturbance pollens make a strong appearance, and grasses fall back; does this indicate arable farming? Pine had gone a little earlier, so this disturbance is Late Bronze Age, and may be related to the walls on the Downland. At 90cm, where the fen-peat gives way to blanket-bog, a stone trackway was laid on the bog surface, and disturbance pollens reappear; this level was given a ^{14}C date of 1720 BP, so Iron Age activity in the vicinity is clearly shown. There is further disturbance at a higher level.

The presence of pine in the top sample suggests that at this point the surface of the bog may be intact.

years ago, and it is tempting to think that alder and oak may have followed different migration routes when returning to Ireland. Oak moved north up the coast from Europe when post-glacial sea-level was still low. Alder failed to follow, and only moved into Ireland from the east after much of the south part of the Irish Sea had been flooded by rising sea-level. From the east coast alder radiated out, taking almost 2000 years to reach the south-west.

Another problem about dating vegetational developments in the south-west is the date at which pine ceased to be an important component in the forests. After growing successfully in most parts of Ireland for a long

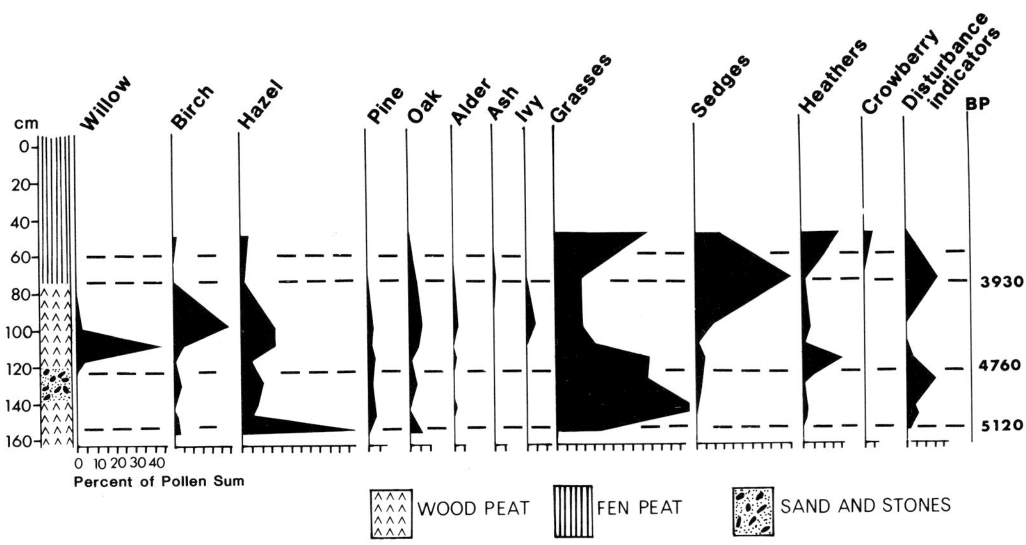

16 *Emlagh Bog Embayment (after Coxon 1985, 37).*
The section here is 160cm deep; pollen samples were counted between 50cm and 155cm.
The lowest peat was a wood-peat, with nearby a basal age of 5120 years BP. The lowest sample had high values for hazel and low values for grass and disturbance pollens. This picture was then reversed; hazel fell dramatically, grasses and disturbance pollens rose and, as at Reenarea, heather made a brief appearance. A layer of sand and small stones was present in the peat, and this was part of a wall dated to 4760 BP. The wall and the clearance phase are Neolithic. As woodland recovered, willow and birch flourished briefly and hazel, oak and ivy regained ground. An oak tree buried in peat nearby had a date of 4600 BP. Fen-peat then became established on top of the wood-peat, and the lower layers of this peat show another clearance phase. Nearby a trackway of small stones lay at this level in the peat, and it was dated to 3930 BP. Clearance and trackway must be of the same age, and indicate Early Bronze Age activity in the area. Blanket-bog must then have started to form nearby, as the pollens typical of blanket-bog (grasses, sedges and heathers) dominate the top sample.

time, pine generally disappeared, though it did hold out in some favoured pockets. Its extinction usually took place about 4000 years ago, and this
11 was the date at Ballinloghig. It held out a little longer at Emlaghlea and in Valencia till about 3000 years ago, but in the Killarney area it did not disappear until 1300 years ago. The later history of the pine in Ireland is a subject of considerable debate.

These datings of the establishment of the alder and the disappearance of the pine are of very considerable importance in Valencia when we come to try to determine the dates at which bog began to form in different parts of the island, and also to introduce chronological levels into pollen diagrams for which ^{14}C dates are not available. It will be assumed that alder became established 5000 years ago, and that pine disappeared 3000 years ago. This date for the immigration of the alder is supported by
12 three pollen diagrams. In the Clynacartan diagram alder makes only

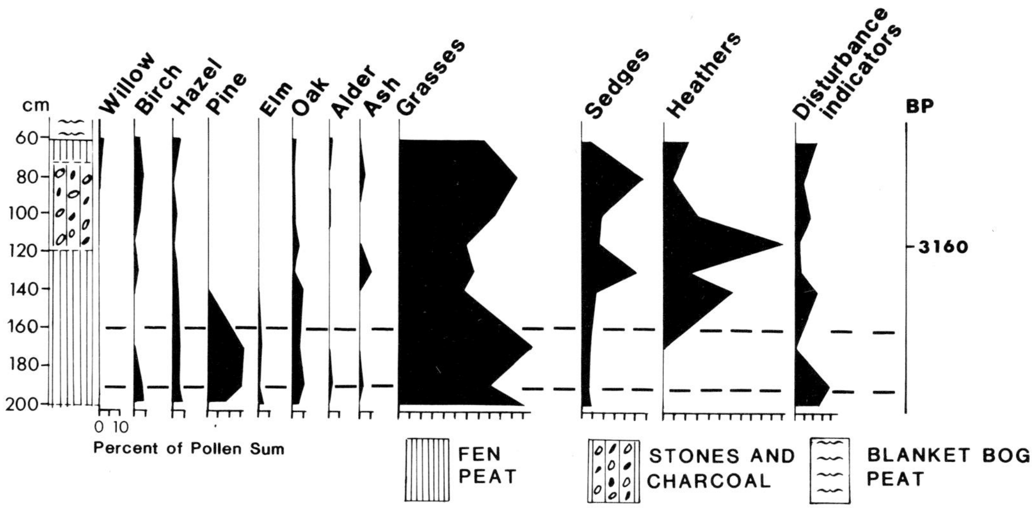

Bray West Bog Embayment (after Coxon 1985, 45). 17
The section here had a depth of 200cm; pollen samples were counted between 60cm and 200cm.

Here there is no basal wood-peat; fen-peat was the first to form. When peat formation started there were, except for some pine and oak, few trees nearby. A clearance phase, probably Neolithic, was in full swing, as the grass and disturbance curves show. Weeds then fall away and pine makes some recovery, but grasses remain dominant. Above 160cm weeds and heathers rise, indicating grazing-land with heather, of Earlier Bronze Age date. Pine then disappears. At 130cm the rise in sedge pollen probably indicates the formation of blanket-bog in the vicinity.

Between 125cm and 75cm charcoal of ling and stones from a small hearth are present in the peat, and the disturbance curve continues, but the tree curves virtually disappear. A sample from the base of this layer had a ^{14}C date of 3160 BP, an Earlier Bronze Age date. At 60cm the fen gives way to blanket-bog.

occasional appearances. In the Emlagh diagram, associated with a wall 16
of Neolithic age, the sample at the base of the peat dated to 5150 years ago had no alder pollen; samples immediately above had very small quantities of alder pollen. At Reenarea Rise alder appeared after 5630 14
BP.

After this time attempts to date the Valencia samples by their tree pollen content run into difficulties. About 5100 years ago there appears to have been a serious outbreak of a disease which killed many elm trees in Ireland, thus bringing about a marked reduction in the amount of elm pollen produced. But the acid soils of Valencia did not favour the elm; pollen amounts in all samples are small, and it is not possible to trace any significant reduction that might be attributed to disease.

Mesolithic man placed a stone platform on the bog at Clynacartan 12
about 6500 years ago, but any alteration he produced in the woodlands was on a limited scale. Neolithic activity, which started about 5000 years ago, initiated a process which ultimately removed almost all trees from the island. Agricultural operations by these first farmers altered the local

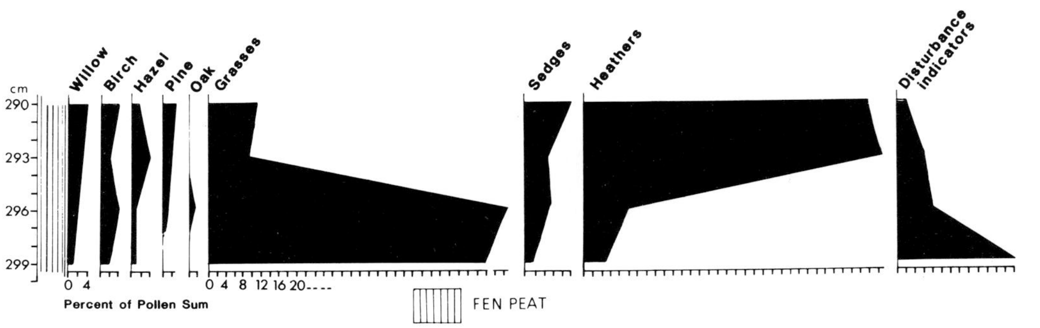

18 *The Imlagh Basin (counts by Peter Foss).*

The bog here was 300cm deep, with 75cm of blanket-bog peat resting on 225cm of Phragmites-*peat, which became very muddy at the base where there was no wood-peat. At a depth of 100cm in the* Phragmites-*peat a horizontal slab of stone, 45cm across and 6cm thick, protruded from the peat face. Peat at the level of the slab had a* ^{14}C *date of 2920 BP, an Earlier Bronze Age date. Counts were made from the basal 10cm only.*

The very short diagram is very similar to that from the Bray West Bog Embayment (about 1km to the west) at a level of about 170cm. The two basal samples show grassland rich in Rosaceae, *possibly* Alchemilla, *which in the two upper samples gives way to heath. As at Bray West, the grassland is probably Neolithic in age; alder is not recorded, suggesting that the peat base here is older than 5000 BP; also a two-thousand-year time-span would not seem excessive to allow for the formation of 200cm of* Phragmites-*peat before slabs were laid on the peat surface about 3000 BP.*

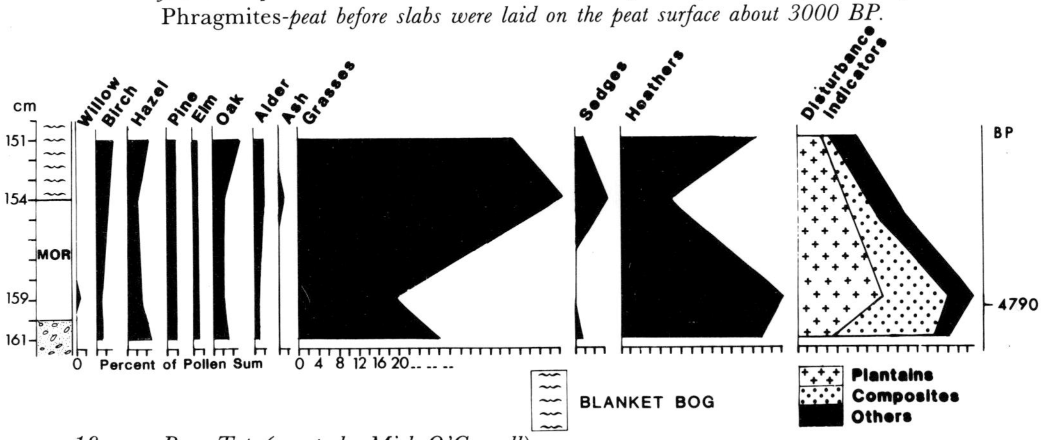

18 *Bray Top (counts by Mick O'Connell).*

The blanket-bog peat was 160cm deep, and rested on mor 6cm thick. Counts were made from the basal 10cm only.

The lowest sample was from the top centimetre of mineral soil. Its pollen content was very similar to that of the sample from the base of the mor. The small amount of various tree pollens included alder, so deposition must have started after 5000 BP. The general picture is of dry grassy heath, the grass being rich in weeds associated with grazing. The sample from the base of the mor was very rich in charcoal fragments and had a ^{14}C *date of 4790 BP. The sample from 154cm shows that grasses and weeds had become dominant over the heathers. The top sample, which, unlike the others, contained no charcoal, shows reduced grazing pressure and re-expansion of heath. It contains pollen of pine and therefore must have accumulated before 3000 BP. The phase of grazing recorded is Neolithic.*

vegetation. At lower levels the removal of trees made it possible to create fields in which plants — both crops and weeds — new to the island were grown. Pollen from the fields was preserved by bog growth and can be *14*
detected in pollen counts, as is clearly seen in the diagrams from Reenarea *15*
Rise, the Brendan Rock-knoll and the Emlagh Bog Embayment, and *16*
probably in the diagrams from the Bray West Bog Embayment and the *17*
Imlagh Basin also. A curious feature of the Valencia counts at these levels *18*
is that they have not shown any cereal pollen.

The short diagram from the top of Bray Head suggests that trees may *18*
have found it impossible to grow on the windy summits of the Tops and on the upper slopes, and grassy heath may have been dominant here for thousands of years. The Neolithic farmers burned the heaths in order to favour the grasses for grazing purposes. By doing so they may have degraded the soil, and so led to the development of blanket-bog on the Tops.

We can see further disturbances in the earlier part of the Bronze Age, from 4000 to 3150 BP, and these also indicate extensive areas of grassland on the island.

The Brendan Rock-knoll shows a marked expansion of disturbance *15*
pollens shortly after the disappearance of the pine at 3000 BP, and this must be Late Bronze Age in date and contemporaneous with the wall system on the nearby Coastal Downland, where a wall was dated to 2680 BP. The same diagram shows a further rise in disturbance pollens at the level of a stone trackway embedded in the peat; the level was dated to 1720 BP, an Iron Age date. Both the continuing development of bog and the activities of successive groups of prehistoric farmers must have brought the numbers of trees growing on the island to a very low level.

The widespread formation of peat

In discussing the origin of the peat that until recently covered so much of Valencia I will draw heavily on Jessen's description of bog formation in Ireland. He described as *soligenous bogs* those in which the formation of peat was controlled by the movement of surface water on sloping ground, from which the water had collected a certain quantity of nutrient salts. The water became sluggish on valley bottoms, fen plants invaded, and their debris began to build up peat. The reed (*Phragmites australis*) flourishes in such conditions and can form great accumulations of peat. Fen-woods with birch and oak and a field-layer of swamp and fen plants can also build up a more or less swampy wood-peat. If the downslope supply of water varies, *Phragmites*-peat and wood-peat may alternate with one another. If the supply of water continues, peat formation extends upslope.

Jessen described as *ombrogenous bogs* those in which the only water available for plant growth was precipitated directly on to the bog surface. One variety of ombrogenous bog is *blanket-bog,* which may entirely cover the surface of the ground, plateaux and slopes alike (where the latter are not too steep), and is developed to the greatest extent in regions with a pronounced oceanic climate. This type of bog develops under almost the same conditions as soligenous bog, and to a great extent it can be regarded as a further impoverished development of a basal soligenous peat. The soligenous peat was fed by drainage water with some nutrient content moving down from slopes not yet covered by peat. As the peat cover gradually extended up the slopes the amount of nutrient available was reduced, and the vegetational pattern changed as plants which could flourish on rainwater only expanded into blanket-bog. In western Ireland purple moor-grass (*Molinia caerulea*) grew very freely under such conditions, and great quantities of *Molinia*-peat gradually built up. Black bog-rush (*Schoenus nigricans*) also contributed to peat formation.

Peat has probably been forming since the disappearance of cold at localities where the local topography caused surface-water drainage to be very sluggish. At Emlaghlea wood-peat with much birch had started to form even before hazel had established itself; this means that the peat base here must be more than 9000 years old. Hazel was already present in the oldest peat that I saw on Valencia, which should therefore be somewhat younger than 9000 years.

I think we can picture that about 8000 years ago the whole of the lower part of Valencia Island was covered by woodland. Along the banks of whatever drainage streams there were, leaf litter and moss debris would slowly raise the level of the ground. The raised banks would occasionally collapse into the stream, impeding the drainage and giving rise to waterlogging and occasional flooding. Fen plants would come in, water-tolerant trees would continue to grow, and the accumulation of swampy wood-peat would begin; the drainage would be further impeded.

In Valencia the oldest record of peat formation is at about 7000 BP,
when muddy *Phragmites*-peat started to form in a reed-swamp in the 12
Clynacartan Bog Embayment. Swampy wood-peat was still beginning
to form around 5250 years ago, because the date of the base at Reenarea 14
Rise was 5630 BP, and that of the base in the Emlagh Bog Embayment 16
was 5210 BP.

At this time the climate of Ireland was enjoying an interlude of warmer summer temperatures and moister atmosphere. Sometimes described as the 'Climatic Optimum', it is thought to have lasted from about 6000 to 4000 years ago. Higher temperatures meant more vigorous plant growth, and moister conditions favoured the accumulation of plant debris, with the result that soligenous peat began to form more and more widely. But the initiation of peat development and its further expansion was not just a response to climatic change; it must also have been closely connected with local topography.

At the base of the bog in one place wood-peat may give way to
Phragmites-peat; elsewhere *Phragmites*-peat gives way to wood-peat, which
may be re-invaded by *Phragmites*. These alternations between wood-peat
and *Phragmites*-peat may also reflect minor oscillations between wetness
and dryness. At the Neolithic site in the Emlagh Bog Embayment, a 16
small oak stump in wood-peat, which was stratigraphically above the
wall (dated to 4760 BP), was dated to 4600 BP. There were further stumps
at the same horizon nearby, and similar stumps were seen in other
cuttings.

Sometimes heavy rains would send a flash-flood of water out into the *Phragmites*-swamp. The flood-water carried along sand and fine grit, and this was deposited as a sandy sheet of debris, later to be buried by peat. Sections through such sheets are occasionally seen in peat-cuttings.

Ultimately, as the soligenous peat thickened and by its own lateral expansion reduced the amount of nutrient carried by water onto its surface, its vegetation had to change as plants of the ombrogenous blanket-bog community invaded its surface. What were these plants? If the vegetation was similar to that of today, it was dominated by *Molinia* and *Schoenus,* had some heathers, especially the ling, *Calluna,* some bog-moss, *Sphagnum,* and bog-asphodel, *Narthecium ossifragum,* together with scattered mosses and lichens. The assemblage had a very low nutrient demand. But the formation of blanket-bog was not confined to the surface of pre-existing soligenous peat; it also invaded slopes that had been hitherto free of peat.

In many places outside the limit of soligenous peat we find at the base of the blanket-bog a distinctive thin (about 10cm) layer; this is greasy and amorphous in texture, black or brown-black in colour owing to a high content of charcoal from local fires, and often rich in seeds of rushes (*Juncus* spp); it is called 'mor'. It rests directly on the mineral soil, which

is often podzolised. In Valencia I noted it below the blanket-bog on the
57 lower slope of the Coastal Downland and on the top of Bray Head; it
18 probably has a much wider distribution. The origin of mor is the subject of much discussion, though it seems to be generally agreed that it coated the ground with an impermeable layer whose wet surface would promote the growth of peat-forming plants, hence the abundance of seeds of rushes. A recent study of an extensive blanket-bog area in Wales showed that mor formed in different places at different times ranging from 7600 to 4000 BP. Two unanswered questions remain: what vegetation preceded the initiation of mor formation, and what was the origin of the fires that produced the charcoal?

Mick O'Connell is working on the first problem in the Connemara National Park. At one point the soil immediately below the mor was rich in varied tree pollens, suggesting woodland; pollen in the mor itself suggested open woodland scrub with birch and holly; pollens of blanket-bog vegetation came crowding in at the base of the peat. Charcoal is abundant in both the mor and the base of the peat. The mor was dated to 3135 BP, a time when Earlier Bronze Age man was active in Ireland. On the other hand, the pollen in the soil immediately below the mor on the top of Bray Head suggested heath interspersed with weedy grass; the date was 4790 BP. Mor on the Coastal Downland was dated to 2900 BP.

The second question is, therefore, 'Were these fires natural outbreaks, or did early man keep areas open by systematic burnings?' If burnings by man contributed to the formation of mor, man will have had a hand in the beginning of peat formation. The formation of mor was not in all cases followed by the development of peat. On Bray South Slope, where soil drainage was good, a field-bank was thrown up on the old ground surface; below the bank there was 6cm of mor resting on a weakly developed podzol in shaly rock-debris. But on the top of Bray Head the fires were almost certainly organised by Neolithic farmers; fine particles of charcoal and ash will have made their way down into the pore-spaces in the soil below, and so impeded the drainage. The ground became waterlogged, and blanket-bog started to form.

The initiation of the formation of blanket-bog in different places was undoubtedly spread out over a long period, and can in some places be
72 shown to be quite late. At the Brendan Rock-knoll some stone crosses were erected, probably not before AD 800. Since they were erected they have been partly buried to a depth of 75cm by blanket-bog peat; on the flanks of the knoll, peat has formed directly on rock.

What does all this add up to in terms of bog formation on Valencia Island? Here there were four types of development:

(i) blanket-bog on the High Tops;
(ii) soligenous peat followed by blanket-bog peat in the Imlagh Basin;
(iii) blanket-bog peat on mor on the Coastal Downland and elsewhere;
(iv) soligenous peat followed by ombrogenous peat on the south-eastern slope of the island.

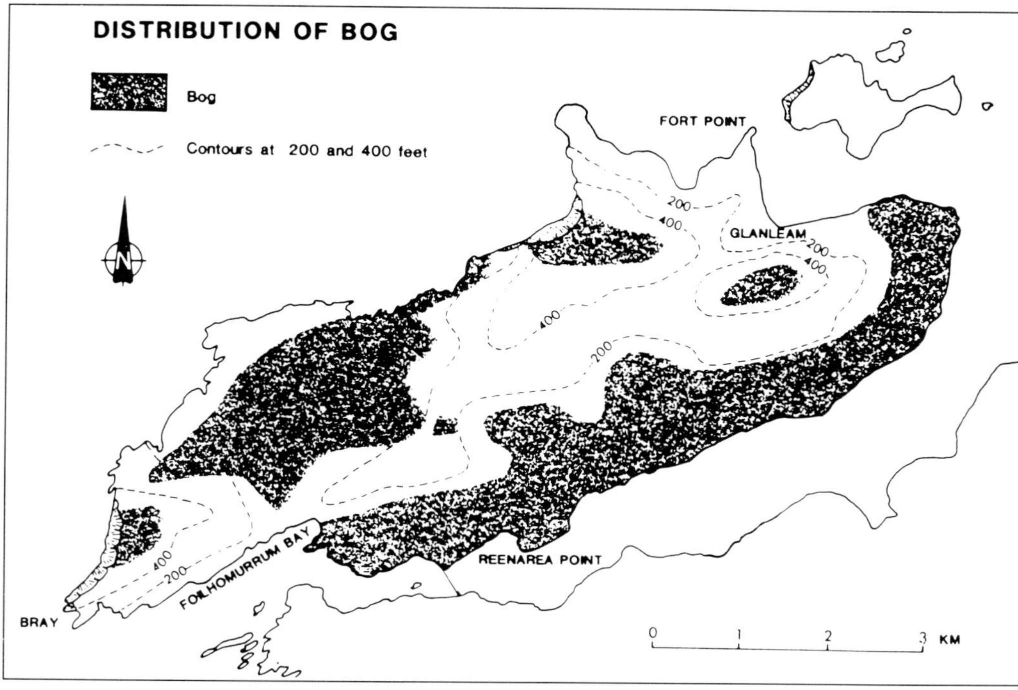

Probable maximum extent of peat before large-scale removal by man began towards the end of the eighteenth century. 23

(i) The High Tops

Lateral movement of water was very poor on the domed surfaces of Bray Top, Feaghmaan Top and Kilmore Top, and blanket-bog built up on the poorly-drained surface. Mor was present at the base of the peat on Bray Top. The peat was not being eroded.

(ii) The Imlagh Basin

I did not see any open-water muds anywhere along the floor of the drainage stream of the basin, and it would seem that early in post-glacial times surface water would have had an unimpeded run to the sea. As noted above, I envisage trees crowding down to the edge of the drainage stream and plant debris accumulating. *Phragmites* would invade the saturated debris and gradually begin to build up fen-peat. Impedance of the main stream would affect its tributaries also, and these began to fill with soligenous peat, gradually changing from stream valleys into bog embayments. Climatic conditions and local topography would determine whether fen-peat or wood-peat would form. In places ponding-up of water would bring about very wet conditions. When a certain thickness of soligenous peat had formed — say 100cm — and run-off no longer reached the bog surface, nutrient supplies would be reduced and ombrogenous vegetation would invade to form blanket-peat. Owing to drainage and cutting a great deal of water has been lost from the ombrogenous peat, but it still occurs to a thickness of 2m in some places; its original thickness will have been very much greater. *Molinia* is the most important constituent of much of the blanket-peat, but some features characteristic of raised bogs could be seen. In a few places unhumified *Sphagnum*-peat forms the top layer. Elsewhere pool and hummock features could be seen. It is very probable that in some areas of the basin the bog rose into the dome-like form of a raised bog. Stokes, writing in 1715, describes some parts of the bog as 'Red Bog', a term used by countryfolk to describe a raised bog. Nimmo in 1814 referred to 'flat red bogs'.

(iii) The Coastal Downland

The Coastal Downland, which is underlain by glacial gravel, slopes south
71 from the north coast of the island down into the Imlagh Basin. On it
a series of parallel walls, built mainly of stones, some of which are quite
large, runs down to the bog in the Basin.

52 The most prominent wall (No. 13), which starts from Coosnaraka,
is about 650m long and ends in the Basin. As it runs down towards the
55 stream, where it is seated on brown muddy peat, it is composed of earth
and small stones; peat with sand washed out of the wall onto the bog
surface had a ^{14}C date of 2680 BP. Immediately to the west of the wall
there is an area (No. 12) of formerly peat-covered small polygonal fields,
56 demarcated by lines of large stones; among the fields there are some
small ruined structures, possibly houses. These walls are seated on mor,
which was dated to 2900 BP. The mor is covered by blanket-bog peat,
57 and the following sequence was noted: 35cm fresh *Molinia*-peat, 45cm
humified moss-peat, 7cm black mor with silt and small stones, bleached surface of mineral soil. Clearly soligenous peat had not formed here. Three borings in the vicinity also showed mor at the base of the peat. One of these borings was through an earth and stone fence that linked up with Wall 13 and was also buried by blanket-bog at its southern end, though this petered out as the fence was followed upslope northwards. Near here Wall 13 must be seated on mor, which gives way to soligenous peat where the wall runs out into the bog. If the contact point could be located, it would show the level that soligenous peat had reached at the time Wall 13 was built, and the level it was at when the wall, and the mor-covered slope above it, began to be buried by blanket-bog peat.

The blanket-bog crept gradually northward up the Downland, and although it did not reach the cliff edge it drastically reduced the amount of ground available for grazing. The quality of the grazing would have fallen also as mor expanded and soil fertility fell.

(iv) The South-east Slope

I believe that there was a very considerable formation of peat on the
23 lower gentle slopes of this part of the island, from the Knightstown Nose
on the east to the Dysert Slope on the west. The peat was soligenous in origin in its basal layers, ombrogenous in origin above, and extended

from the shore uphill to the break in slope from steep above to more gentle below.

As early as 1830 enough peat had been removed from the tip of the Knightstown Nose for a small settlement to spring up there, called appropriately enough 'The Nose'. The lower road along the shore from Knightstown to Chapeltown did not exist when the first six-inch OS map was published in 1846, and the base of the slope may still have been encumbered by peat. Peat still remains on the Reenarea Rise, on the lower slopes of the Oonagranshee Lobe north of it, and in the Donnybrook and Chapeltown valleys. But the bulk of it disappeared as population rose in the early part of the nineteenth century. There was a high premium on peat-cutting. As peat was cut away, new land was revealed. Any bog-timber encountered could be used for building or as fuel. The cut peat could be carried by boat to Cahersiveen or Dingle and sold for cash, while on the return trip calcareous gravel and sand might be collected from the beaches on Beginish and used to fertilise the podzolic soil below the bog.

Man's reaction to physiography and to the changing environment

In all I have defined thirty-five physiographic units (see map at end), and I have chosen an arbitrary order in which to describe them and the antiquities they contain. The units are shown on a map, which has attached to it two lists of the unit names: in one the names are in alphabetical order, and in the other in numerical order following the sequence of the descriptions. I take first the north end of the island. I then move westwards down the spine-ridge of the island, and on to Bray Head. I turn north around the perimeter of the Imlagh Basin and then enter the Basin itself. I exit from the Basin by its southern arm at Clynacartan, and follow the south-east slope of the island to Knightstown, with a final look at the shelf nearby on which the Kilmore cemetery is situated. Some unit names will appear before that unit has been described, but reference to the map on the endpapers will make location easy. The area in hectares given for each unit is no more than a reasonable approximation.

The principal townland or townlands in which the units lie are given; where a unit sprawls across several townlands, the two in which it occupies the largest area are stated.

In each entry the physiography is described in the opening paragraph. Description of the antiquities (if any) present in the unit follows.

Where possible, I place the antiquities noted in the following periods, using uncalibrated ^{14}C dates (where appropriate) for approximate datings of the opening of the period:

Mesolithic	9000 BP
Neolithic	6000 BP
Earlier Bronze Age	4000 BP
Late Bronze Age	2700 BP
Iron Age (say 450 BC)	2400 BP
Early Christian period	400 AD
Historic period	1200 AD
Period of population growth	1750 AD
Period of readjustment	1850 AD

THE NORTH END

35 **1 Dohilla Slope 105ha Dohilla, Feaghmaan East and West**

The main slope is to the north-east, convex, steep and rough down to 60m, flatter and smooth with good soils below that level. The west side

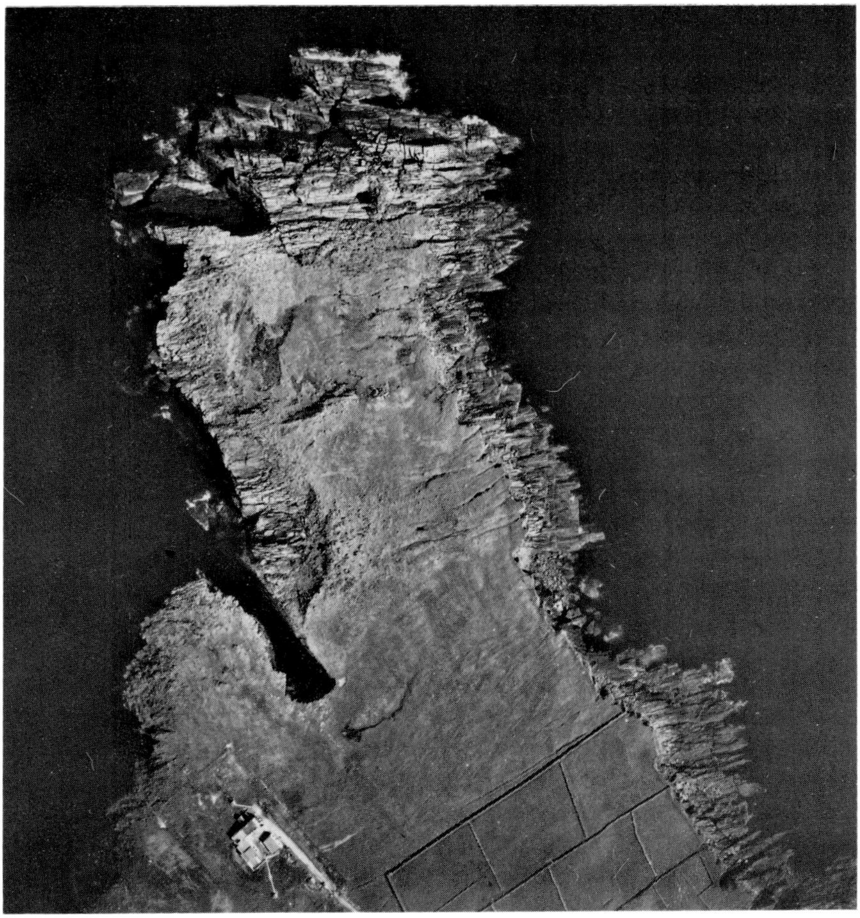

Dohilla Slope, Reenadrolaun Point. 27
Here at the north tip of the island the north-east/south-west trend of the rock structure is well seen. The chasm north of the radio station is also structurally controlled. The grassy slope east of the station still carries early field walls; some traces of these still survive in the enclosed fields to the south.
Based on Ordnance Survey by permission of the government (permit 5149).

of the northern end of the Slope merges into the eastern end of the Fogher
cliffs. Reenadrolaun Point is exposed to very violent wave attack. There 2
are some dunes of very coarse sand at the base of the Point. Thick peat 27
never formed in this part of the island, possibly because of the shape of the ground, or because of a rain-shadow effect from the high ground to the south-west.

There are several old stone walls on the upper slope, and the big
nineteenth-century slate-quarry was cut into it. Most of the lower slope 35
is laid out in regular fields, but east of the radio station surviving lines of boulders mark out an old field system. On the low exposed ground leading to the Point a well-built animal shelter has been inserted in a

fissure in the rock. Inshore from Coosadillisk there are low, crescentic animal shelters of the type described by Brian O'Kelly on Beginish. I did not see the sheepfold shown on the six-inch OS map.

2 **2 Fort Point 65ha Glanleam, Gortgower**

This is a slope formed of stepped ground. On the south it has relatively
108 flat improved ground at about 70m. There is then a rough stepped slope
of rocky ground 200m wide, and then low-lying (below 15m) improved land.

On the rocky slope, 150m west of a nineteenth-century memorial cross, there is a ruined oval house, measuring 3m x 2m internally; there is a suggestion of an annexe at the northern end. The house lies on a small terrace and there is some revetting nearby. An Early Christian context is possible. The low-lying land is laid out in regular fields, except near the Point itself where there is some rocky ground. Here there are the remains of a settlement but these have been greatly disturbed by reclamation. There is a built-up rectangular house platform, 5m x 3m, two circular enclosures (corrals or fields?), and remains of a small walled area.

The tip of the Point carries a seventeenth-century star-shaped fort and a modern lighthouse. A gallaun stands within the fort.

3 Glanleam Pocket 65ha Glanleam

This deep, steep-sided small hollow is tectonic in origin. In the pocket
wind-flow is reduced and freezing temperatures are rare, and today sub-
35 tropical trees grow freely in it. The kidney saxifrage (*Saxifraga hirsuta*),
common in north-west Iberia but very rare in Ireland, grows luxuriantly on the steep south wall. One is tempted to think that here perhaps some plants and animals might have found refuge during the colder periods of the not too distant past. But the north-east aspect of the Pocket would make it a suitable trap for snow, and it is not impossible that a semi-permanent snow-bed lay in it during the last advance of ice in Ireland. On the northern slope of Foilclogh, 23km to the east on the mainland above Cahersiveen, there was just such a permanent snow-bed at 250m. The shape of the Pocket almost suggests a corrie of former times.

This sheltered hollow ought to have been a most attractive location
108 for early man. But it has been landscaped and planted for at least one
hundred years, and no early antiquities have been noted.

On the west, where the top of the slope is beginning to flatten out, there is a Children's Burial Ground surrounded by a wall. The interior is completely overgrown by dense impenetrable scrub. Nearby is a holy well, *Toberfinnan*. This has been reconstructed, and has a modern slab of slate with an incised human figure. It is still venerated.

An old road to the west formerly ran along the floor of the Pocket. About 100m from the point where the modern road to Glanleam House leaves the old road there is a small plain cross.

Kilbeg Top, ruined wedge tomb. 29
This ruined wedge tomb lies at the lower margin of the Kilbeg Top. Its siting is analogous to that of the well-preserved wedge tomb on the Kildreenagh Slope. It looks south-westwards over ground that drops away to the Portmagee Channel.

4 Kilbeg Top 100ha Kilbeg East and West, etc. 33

This is a gently oval dome, rising to 190m, cut off to the north by the abrupt slope down into the Glanleam Pocket and less sharply to the east by the slope down on to the Kilmore Shelf.

The Top is crossed by many walls, generally running north/south and ending abruptly on the north when they meet the rim of the Glanleam Pocket. Several of the walls mark townland boundaries.

Just above the southern boundary of the Top, at about 150m, there is a very ruined wedge tomb. One side slab, 2.50m long and oriented 29
north-east/south-west, still stands; there are two large prostrate slabs nearby. Its elevation and aspect are very similar to those of the well-known wedge tomb on the Kildreenagh Slope. It is known locally as 35
'The Piper's Stones'. A gently sinuous heavy earth bank (crossed by later field boundaries), lying east/west, runs past it a few metres to the south.

The immediate surroundings are covered by later cultivation ridges.

5 Gortgower Saddle 45ha Gortgower, etc.

From the Top we move across the Saddle, a broad col at about 100m leading across to Geokaun Top. On the south-facing slope of the Saddle soils are good. There are important modern road junctions on the Saddle, connecting the north tip with the south-western parts of the island.

Just where the Saddle is rising into Geokaun Top there is a ruined settlement which may be Early Christian in age. The ruins of an oval mound, 10m across, contain an oval drystone house, 6m x 4m internally; there are some big erect slabs nearby. Some 100m to the south-west are the ruins of a cashel, 30m in diameter, with stone walling visible in places. A short distance to the west some blocks placed against a rock outcrop suggest the ruins of a small hut, and there is also an animal shelter, 3m x 1.5m, very like those described by Brian O'Kelly on Beginish.

At about the same level, 200m to the south-west, there is a cashel, which was later altered, perhaps in the nineteenth century. It measures about 30m in diameter and has some very large slabs in its bank. Much of the east side of the bank has been removed, and there are the rectangular earth foundations of two small huts inside the cashel; a third lies outside the bank on the south-west. A short straight length of high heavy bank has been built on top of the northern section of the cashel bank, possibly as a wind-break. There is a small featureless circular bank, about 25m in diameter, 150m downslope to the south.

South again, where the Saddle is giving way to the Kildreenagh Slope,
114 is the upslope end of a long straggling clachan, to be described later.

All these features lie on the south-facing slope of the Saddle where there are well-drained good soils; I did not see any antiquities on the north-facing slope.

54 **6 Geokaun Top 75ha Feaghmaan East and West, Tinnies Upper**

The Top is an oval dome which forms the highest point of the island at 295m. The summit is covered by blanket-bog; the slopes have rough pasture. On the western side, as we approach the Fogher cliffs, gullies
54 formed by incipient cliff collapse would have afforded good shelter for stock.

A short distance north-east of the summit and high above the old slate-quarry are the frost-shattered remains of a tor. A narrow terrace at the foot of the tumbled rocks holds the remains of a settlement. There is the outline in stone flags of a house, 10m x 5m, and nearby two small rectangular outlines, perhaps turf-drying racks. Tucked in among the tumbled boulders of the former tor is a well-built animal house (3m in diameter) with an outer wall of large blocks; there are others nearby. A wall running up from the slope below curves around below the settlement. Such an exposed site facing north-east at 250m would seem to be a most inhospitable place for farming. Could it have been an observation post overlooking Valencia Harbour? Its age and function remain unknown.

7 Fogher Mega-cliff Feaghmaan East and West

54 These high cliffs rise to almost 240m on the western side of Geokaun Top. They were cut across a lobe of higher ground which formerly extended further to the west. The cliffs are quite unstable, and heavy rock-falls occur from time to time. Gullies, some quite large, parallel the cliff edge and indicate the sites of future falls.

THE SPINE-RIDGE

8 Kildreenagh Dip-slope 115ha Tinnies Upper, Cool East and West

On the upper south side of the Spine-ridge that runs south-west down the island there is a very obtuse triangular facet, sloping south-south- 35
east above 100m. From its apex, at 225m on the shoulder of Geokaun Top, the long west side runs down along the Spine-ridge till the Slope comes to an end at 100m, breaking up into minor parallel ridges as it approaches the Coarhamore Saddle. The short east side drops down to 135m in the Gortgower Saddle. The base is 2.5km long, and drops from 135m in the east to 100m in the west. Towards the south-west the Slope is interrupted by the upper end of the Tinnies Lobe, which falls away to the south-east. For human occupation this slope shares with the south slope of Bray Head the distinction of being the most favoured area in the island. The ground faces south, is protected from north and east winds, and the slope, though not steep, is sufficient to inhibit the growth of peat. The soil is deep and free-draining, coming into the category of brown earth or brown podzolic.

There is a splendid Early Bronze Age wedge tomb at about 400m
towards the western end of the Slope. Like the other wedge tomb in 35
Kilbeg Top it is oriented towards the south-west, and commands an open view of the Portmagee Channel. It is situated where rough ground above is giving way to better land below.

The Slope was popular in Early Christian times. We have already noted the settlement on the Gortgower Saddle immediately to the east. On the Slope itself there are four sites and several gallauns to be noted, all lying close to the road which traverses the Slope. The route followed by the road must go back at least to this period.

Moving down along the road from the east we have on its upslope side a poorly preserved site, probably originally ecclesiastical but later taken over for a burial ground, traditionally for unbaptised children; it is marked *Calluragh* on the six-inch OS map. It contains very delapidated ruins of former buildings — one possibly a tomb-shrine — obscured by many small stones serving as grave-markers.

On the downslope side is a gallaun.

Next on the upslope side is a cashel, 30m in diameter, with foundations
for structures inside and a gallaun at its entrance. In a sloping field 100m 32
to the north-east are two rectangular house foundations, one with a souterrain; there is another souterrain nearby. These features may represent the remains of a lay village.

There is then on the upslope side Kildreenagh, clearly an early
ecclesiastical site enclosed by a stone wall. Within there is an ogham 72
stone, to which a cross was added later, and the foundations of several small drystone buildings, two rectangular and three circular. The site was later used for burials, and is packed with small upright grave-markers.

32 *Kildreenagh Dip-slope, lay village?*
Bottom left we see a cashel which has a gallaun at its entrance and the foundations of a building inside. Two fields to the right, a semicircular tractor track leads uphill from a cottage. Within the arc of the track there are two rectangular house foundations, one with a souterrain; there is a second souterrain nearby. Have we here a lay village, with the cashel giving protection to other undefended houses nearby? The date might be about AD 1000.

An ogham stone with a 'Quenga' inscription which formerly stood a little to the west has disappeared.

Beyond the crossroads on the downslope side is a very ruined cashel, and beyond it a gallaun; some scores on the stone are possibly man-made, and may once have been part of an ogham inscription.

At this last site there are also a rectangular house base and a 'garden' (a very small embanked rectangular area), perhaps of the nineteenth century.

There are two clachans on the Slope: one, at the extreme east of the
114 Slope, is a continuation of the clachan already noted on the Gortgower Saddle, and the second, partly on the Slope and partly on the upper end of the Tinnies Lobe, is traversed by the road.

36 9 Tinnies Scarp-slope 35ha Tinnies Upper

On the upper north side of the Spine-ridge a scarp-slope of rock, shaped like a parallelogram, runs north-west down to the sea at an angle of about
33 10°; it ends in cliffs 60m high. In this area the primary bedding of the rock is nearly horizontal, but its strong cleavage is against the slope of
54 the ground and this gives rise to numerous rock overhangs. The slope is bisected diagonally by the townland boundary between Tinnies Upper

and Cool East townlands. The same boundary separates my Tinnies Slope from my Cool Slope.

About half-way along the diagonal at a height of 100m two low rounded bosses of rock stand up, with an area of scattered blocks downslope from 36
them; I believe that in pre-glacial times there were residual hillslope tors here. Ice formerly pressed heavily against the Slope, and a slab of living rock carries striae running north/south. Later frost action moved unconsolidated material (including rock blocks from the tors) downslope, and much of this material went seawards over the cliff edge, leaving very little soil on the Slope. The slope is too steep for general bog formation, but one small flatter area has peat developed on it.

Four linear walls (Nos 7-10) of large blocks run down the Tinnies Slope; there are many small scarp-faces on the Slope, and advantage 34
was taken of these in wall-building, where the face of the scarp could serve the purpose of a wall. I did not see any cultivation ridges or any ruins with mortared walls on the Slope.

Tinnies and Cool Scarp-slopes looking to Geokaun Top. 33
In the foreground we see the dissected end of the Cool East Lobe; Doonroe promontory fort is on the third headland from the bottom. In the centre we have first the stepped Cool Slope, and behind this the uninterrupted Tinnies Slope. In the background we have Geokaun Top, deeply cut into on the left by the Fogher Mega-cliffs.

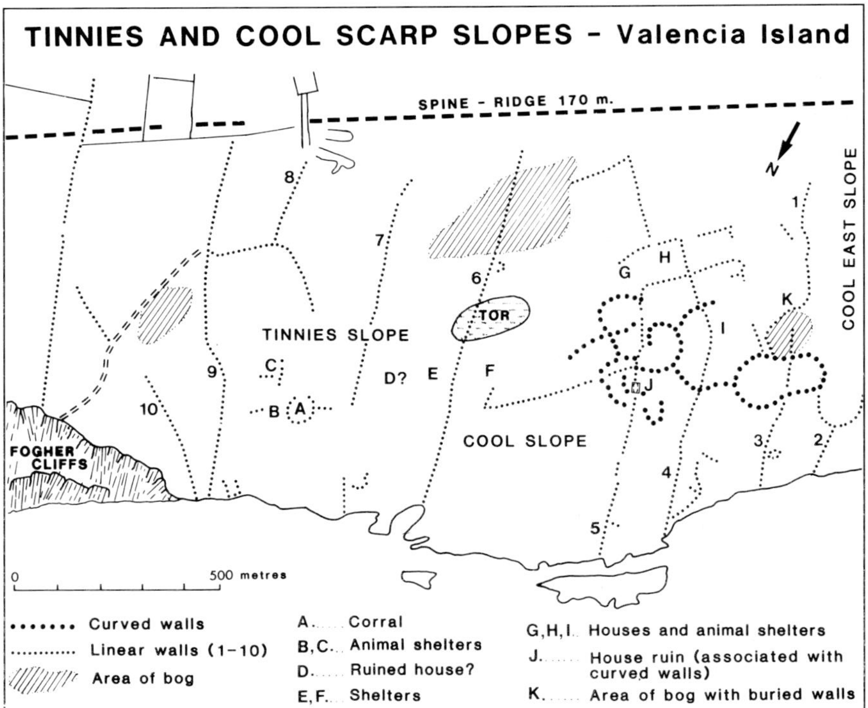

34 *Tinnies and Cool Scarp-slopes.*

36 *In essence these two scarp-slopes are a single slope running down north-west from the Spine-ridge at 170m to cliffs 60m high on the coast. There is very little soil, and only thin peat. The slopes are divided by about ten sub-parallel linear walls of large stones which run downslope; 6, which runs first through a patch of bog and then across the stumps of earlier tors, forms the townland boundary between Cool and Tinnies.*

In Tinnies, apart from the linear walls, signs of human activity are scarce. At A there is a small corral, with some ruined masonry (B) and some animal shelters (C) nearby. Below the ruined tors there are some animal shelters (E) and possibly a ruined house (D?). The shelters continue across the townland border into Cool (F).

In Tinnies the downslope is relatively continuous, but in Cool it is interrupted by flatter treads, with some soil but no evidence of cultivation.

Linear walls 4 and 5 are linked by cross-walls at their upper ends, and there are houses and animal shelters here (G, H, I). But underneath the linear walls there appears to be a series of curved walls also of large stones which enclose smaller round fields. A house ruin (J) appears to be associated with these walls. There is a small area of bog (K) which buries walls and other features, but these have not been dated.

At one point there was a circular enclosure (A), about 30m in diameter. The surrounding wall was too slight for it to be claimed as a cashel, and it may have been a corral for stock. In the interior there is more soil on the downslope than the upslope, and this movement may have been brought about by 'poaching' by confined animals. A small hut-like structure lay against its western side. Nearby a small flat (B) bounded on two sides by low rock scarps shows some traces of crude masonry,

The North-east Slope.
On the left we see Kilbeg Top, with the Kilmore Shelf below it. Behind we see Geokaun 35
Top and the Dohilla Slope (with the big quarry cut into it) running down from it. Below, we see the wooded slope of the Glanleam Pocket. In the foreground we see Knightstown, and to the left the extensive buildings of the cable station, which lie on the solifluction slope of the Knightstown Nose.

West end of island, panoramic view. 35
The wedge tomb stands near the top of the Kildreenagh Dip-slope. A short distance beyond the tomb we see a rock-knoll on the west end of the Spine-ridge. In the middle distance on the left the ground drops into the Coarhamore Saddle, and then rises into the Coarha Top. In the distance behind the Top we see the Skelligs. In the right background we see Bray Head rising behind the Clynacartan Bog Embayment. The dip-slope is to the left, and the scarp-slope to the right.

36 *Cool Slope, prehistoric walls.*
John Neill, then in his eighties, guided me to these walls. Seventy-five years ago his father, also John, had guided Thomas Westropp across the same slopes. The terraced nature of the Cool Slope can be seen behind John's head, and a little higher up the stumps of the degraded tors. Beyond the tors we see the more continuous Tinnies Slope. In the background is Geokaun Top, with the Fogher Mega-cliffs on the left.

36 *Cool Slope, complex of stone walls.*
In this picture, taken in early spring, the wall complex stands out clearly in the dead vegetation. A prominent wall (No. 5) running up the slope divides the picture into right and left halves. At the bottom of the picture a wall from the left meets the central wall; immediately above the junction the ruins of a house (J) lie athwart the central wall. In the centre of the picture the wall cuts through an older subrectangular field. Higher in the picture on the left a circular corral is seen; on its right-hand side its wall shows a gap, perhaps because its stones have been carried off to build the central wall. Running up from the bottom left-hand corner a straight wall runs diagonally across an older field to join the central wall. There are two field systems of different ages here.

Tinnies Slope, animal shelter against rock overhang. 37
We have on the left an overhang with a space beneath. A line of slabs has been erected about 1m west of the lip of the overhang to create a pen (E) for animals. A tape-measure lies across the entrance.

and there may have been some settlement here. A few rough animal shelters (C) also occur in the vicinity.

The tumbled blocks of rock below the tors had substantial voids between them, and some of these had been turned into pseudo-souterrains by stopping the gaps underneath the blocks with drystone walling. Where roofing-blocks were absent but there was a flat partly surrounded by blocks, a short length of wall would create a pen. There were many of these crude animal shelters in the vicinity of the tors (E, F), and at one 37
point there was a suggestion of an outline of a rectangular house (D).

10 Cool Scarp-slope 35ha Cool East 36

With the cleavage of the rock against the slope of the ground, there were numerous small rock overhangs. On its lower, seaward part, the Slope 33
is interrupted by some flatter terraces, and there is a broader flat above the cliffs. Small areas of bog had built up on two of the terraces; one of the bogs had covered older walls. But peat was generally less than 50cm thick, and rested on the leached horizon of a peaty gleyed podzol. There was no evidence of former cultivation.

There was a much greater density of abandoned stone walls on this 34
Slope than on the Tinnies Slope, perhaps because the flatter terraces offered better grazing.

There are two types of stone wall, one curved and enclosing a relatively small area, the other linear and coaxial, running in at right angles from the coast. Of the latter type I identified six walls; there may be more.

38 *Cool Scarp-slope, ruins of house?*
A linear arrangement of large slabs, lying transverse to the hillslope, may represent the last vestiges of a drystone house (J). There are numerous stone walls in the vicinity, and many of the stones of the house may have been carried off to provide wall material.

34 The modern boundary between the townlands coincides with Wall 6.

Two linear walls (Nos 4 and 5) clearly cut across the smaller enclosures. I interpret this as indicating that the curved walls belong to an older agricultural system, and that the linear walls which intersect them are younger. But as there is very little bog on the Slope, it has not yet been possible to get suitable material for ^{14}C age determination. A small area of bog (K) on the west of the Slope does bury a wall which belongs to the linear system, and there is also beneath the peat a partly exposed mass of stones, which might be a pile of field stones or possibly a collapsed hut. The bog ends against a low rock-scarp, and there is some walling associated with the rock face. Some linear walls are connected by short lengths of wall at right angles to them, thus creating small rectangular fields.

The numerous rock overhangs on the Slope offer some shelter. The judicious addition of a few boulders could quickly provide a rough fold or sty, or small walls could give a more sophisticated structure. In the latter case it is difficult to say whether it is an adequate farm building or a rather small house. A number of both types of structure (G, H, I) are scattered over the Slope.

36 At one point there appeared to be the ruins of a free-standing house
38 (J). There was an oval pile of stones (about 8m x 4m) with some larger sub-horizontal long stones which might have been roofing-slabs. It lay in a length of linear wall (No. 5), and it was impossible to decide whether the wall had abutted against a complete house, i.e. was contemporary with it, or whether an earlier house had been a source of stone for a later wall, the wall running through the house so as to shorten the distance

Cool East Lobe, ruined clochauns.
The zigzag course of the central wall follows the ruined walls of some conjoined 39
clochauns. Right of the wall are the conjoined earth and stone foundations of two houses, 54
perhaps later. Left of the wall a rectangular enclosure is suggested by a 'ghost' outline.

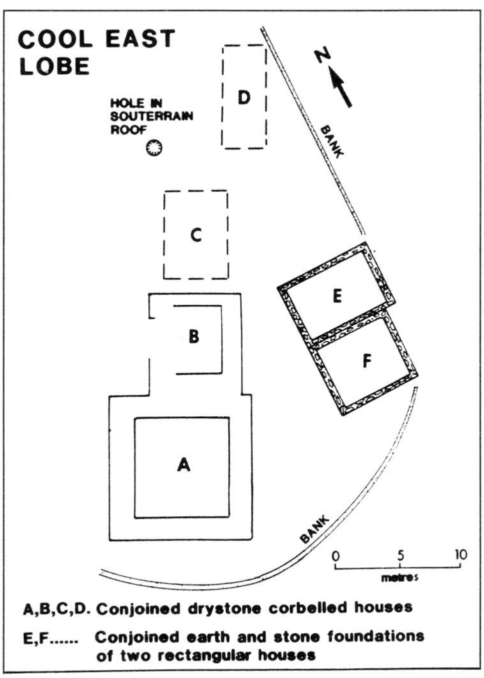

Clochauns. 39
At this point a field boundary wall on the six-inch OS map showed an extraordinary series of kinks. Inspection of the site revealed that the field wall had been constructed on top of the surviving walls of a group of conjoined drystone corbelled houses or clochauns. There had been four such houses, with a souterrain nearby. Immediately south-east of the group were the conjoined earth and stone foundations of two rectangular houses; these may be of later date.

that stone had to be carried for wall-building. At another point where the same wall (No. 5) passed close to a curved wall the latter was very much reduced in thickness, suggesting that stones had been robbed from it to build the linear wall. Thus what slender evidence there is suggests that the curved walls are older than the linear walls.

4 **11 Cool East Lobe 60ha Cool East**

54 The Cool East Lobe is best regarded as a small hogback ridge running
down north-westwards from the southern part of the Spine-ridge; its
seaward end has been very much dissected by wave attack. It first drops
61 down steeply from 160m to 80m, and then develops a relatively flat well-
drained nose, with good soils; below 50m it dips gently to the peat-covered
cliff edge. This was an attractive area for settlement.

The flatter part of the Lobe, between 80m and 50m, is closely covered by small irregular stone-walled fields; though today given over to grazing, many of them show former cultivation ridges. There are several small abandoned farmhouses.

One field wall cuts across a low oval mass of stone debris containing
39 the ruins of three contiguous, well-built, corbelled drystone houses or
clochauns, resembling the well-known examples on Bray Head. There is a fourth house, and an aperture in the roof of a souterrain nearby. One house had its door in the north-west wall. Like those on Bray Head, the buildings do not appear to be churches but rather domestic buildings. On the east side of the mound low earth and stone banks outline two contiguous rectangular houses; these are probably younger than the stone houses. A short distance to the north are the earthen banks of a rectangular cottage; this may be nineteenth-century in date. There are also old curved earthen field banks nearby.

Some 100m to the west are the ruins of a circular house platform, 7m in diameter, with a souterrain below.

54 On the cliff edge, 400m to the north-west, are the remains of Doonroe
promontory fort, described by Westropp. A large double bank and ditch, perhaps unfinished, cut off a headland; inside are the outlines of two conjoined sunken huts.

The rounded top of this Lobe, with its well-drained soils, must always have been an attractive place for settlement, but because of later occupation and disturbance I did not find any evidence of prehistoric activity.

4 **12 Cool West Lobe 85ha Cool West**

54 I find it very difficult to define this Lobe which projects into the Imlagh
61 Basin. Its rock foundation is irregular in outline and variable in altitude;
as it was progressively buried by peat, one part — which I call the Main Escarpment — first became isolated as an island, and then disappeared under peat. Peat-cutting and drainage have since exhumed not only it, but also other parts of the Lobe. Where do I place the boundary? The profile of the Lobe can be crudely likened to that of the cut-off breast

and head of a duck, though the duck has a flat-topped head and an exaggeratedly long beak. The Lobe springs from the western end of the Spine-ridge, where its base abuts against the Cool East Lobe. This south-
eastern higher (about 70m) part of the Lobe is the duck's breast, with 54
the neck and head sloping down to the north-west into the Basin. The 61
duck's head has a flat top, where it ends in a low rock escarpment at about 30m. Between the back of the head and the Cool East Lobe there is lower ground, first drained by a stream running north to Coosheen-nadagallaun, and later filled with peat to form the Cool Bog Embayment. The duck's cheek is at a low level, but the elongated beak — the Main Escarpment — which runs north-east/south-west is higher. There were springs in the vicinity of the cheek, and peat formed quickly here in the Emlagh Bog Embayment, cutting off the beak as an island and making it a favoured site for early settlement. On the higher breast of the duck good soils developed on the rocky substrate. The downslope end of the Lobe (the duck's neck) became largely peat-covered, but there has been extensive land reclamation, with regular fields and access roads. In such an area it is hard to define the original extent of the peat, and hence the margin of the Lobe.

At the north end of the Lobe, a low escarpment forms the flat top to the duck's head. The ruins of a circular hut, about 3m in diameter, stand on the edge of the escarpment, with a souterrain below it, tucked into the face of the scarp. Limpet shells and other debris had fallen through the floor of the hut into the passage below; the shells had a ^{14}C date of 930 BP. The face of the scarp showed signs of revetment, and there were the ruins of a small box-like structure beside the hut. Westropp illustrates the site and describes it as a fort.

13 Coarhamore Saddle 10ha Coarha More 35

The Spine-ridge ends in parallel small ridges and hollows at about 80m. There is then the small Coarhamore Saddle at about 70m, with the ground dropping away to north and south but rising again westwards into the Coarha Top. The relatively horizontal surface of the Saddle is poorly drained, and blanket-bog peat formed here to a thickness of at least 1m. Below the peat there is a compact silty soil of poor permeability, which could be classed as a gleyed peaty podzol.

There are no signs of any early human activity.

14 Coarha Top 65ha Coarha Beg, More 35

It is tempting to regard this Top as part of the Spine-ridge which has been dislocated southwards and elevated. It has the same ridges and hollows with the same trend, and a ridge running downwards from its western end effectively continues the Spine-ridge. The Top rises to 100m, with three minor rock ridges separated by hollows in which peat (now largely removed) formed. The north face is a small steep scarp-slope, with overhanging outcrops and small flat terraces.

On the north face some animal shelters have been tucked in among the outcrops, and the edges of some of the terraces show crude revetting with large stones.

On the south side of each of the rock ridges on the Top are the ruins
114 of a clachan.

There is a gallaun near the base of the more gentle southern slope.

4 **15 Coarhabeg Lobe 35ha Coarha Beg**

From the west end of the Top a broad lobe projects north at a lower
61 level into the Imlagh Basin. The soil on the Lobe is well drained.

There are two complex house clusters on the Lobe, both with souterrains. To the east — 400m north-west of the Sugrue farm — there
42 was the foundation-bank of a polygonal house, about 8m across, with an entrance at the south-east. There appeared to be an annexe at the opposite side of the house, abutting against a short length of curved bank, perhaps the last remains of an enclosure. Underneath the bank, at the point where the annexe met it, was a curved souterrain. It appeared to have been erected in a shallow trench with drystone walling and a capstone roof; it was then concealed by completing the building of the bank. The remains of a circular hut lay a few metres south of the house. A few metres south-east of the house was the earth and stone foundation of a rectangular house, 12m x 8m; a block of quartz lay at its central

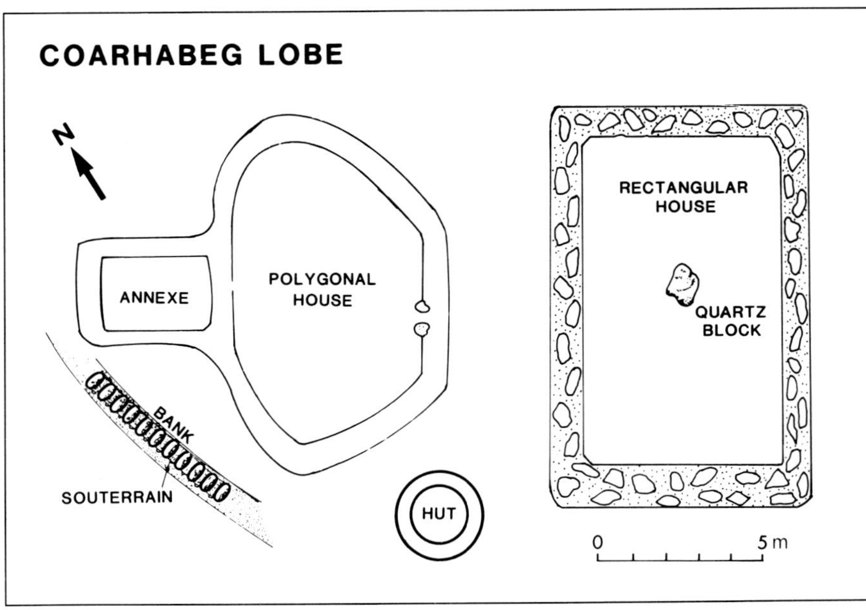

42 *Eastern house cluster.*
There had been a settlement here, which perhaps existed for some considerable time. There was the outline of a polygonal house with an annexe to the rere. The foundation for a circular hut lay nearby. South-west of the house there was a curved bank, and a souterrain had been inserted into this rather than dug deeply underground. To the east were the foundations of a large rectangular house; at its centre there was a large block of quartz.

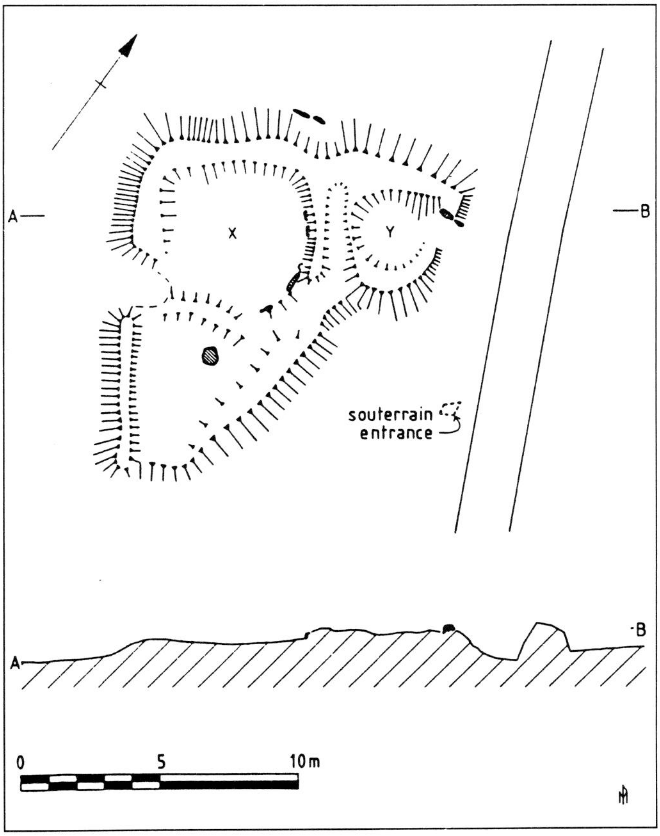

Coarhabeg Lobe, western house cluster. There was a suggestion of a small polygonal house (X), with a smaller rectangular house (Y) to the north-east; also fragments of banks of earth and stone. There was a small souterrain. *43*

point; this building is probably younger than the other structures on the site.

The west site — 500m south-west of the first — was less well preserved and difficult to interpret. There was a suggestion of a polygonal house, with a small rectangular house immediately to the east; there were also fragments of fences of earth and stone. There was a souterrain with a rectangular chamber at its eastern end.

Some 800m north-north-west of the Sugrue farm on a rocky knoll overlooking the Imlagh Basin there was a small house foundation, perhaps the remains of a nineteenth-century 'cabin'.

16 Clynacartan Saddle 10ha Bray

From the west end of Coarha Top the continuing rock ridge runs down into this low Saddle. The Saddle is cut into by cliffs on the south, and on the north slopes down below the peat in the Imlagh Basin. To the *35*
west lies Bray Head. The Saddle forms an important routeway between Bray and the rest of the island.

Where the Saddle rises towards Bray Head there is an isolated souterrain.

There is a clachan and a pretty corbelled well on the Saddle.

44 *Bray South Dip-slope, clochauns and farm.*
On the right we see the track to the signal tower on Bray Head cutting across an early farm. Recent land divisions below the track and at right angles to it cut across the boundaries of the older fields. On the left we see the group of drystone buildings (clochauns), probably of the Early Christian period. Above the track many old boundaries run up the hillside. The straight line of the cliffed coast follows the trend of the underlying rock structure. The protuberance on the coast is named Doonroe Cliff; a fort, later carried away by coastal erosion, may once have stood here.
Based on Ordnance Survey by permission of the government (permit 5149).

BRAY HEAD

35 **17 Bray East Slope 30ha Bray**
This is a straight, steep (about 15°), generally rocky slope, with some
61 hillslope tors; it is the east face of the Bray Horst. It runs east-north-
east down into the Imlagh Basin. The soil is well drained.

The Slope has been subdivided many times by walls and banks running down it. The most recent division was carried out by the Knight of Kerry in the 1870s. Some walls had had a few very large stones included in them; if the wall was subsequently demolished, such large stones were occasionally left standing, giving them the appearance of pseudo-gallauns.

35 **18 Bray South Dip-slope 130ha Bray**
61 This is a long concave dip-slope, facing south-east and ending in cliffs about 30m high. It is the south face of the Bray Horst. The upper slope is steep (about 15°), and is in rock; the lower slope is shallower (about

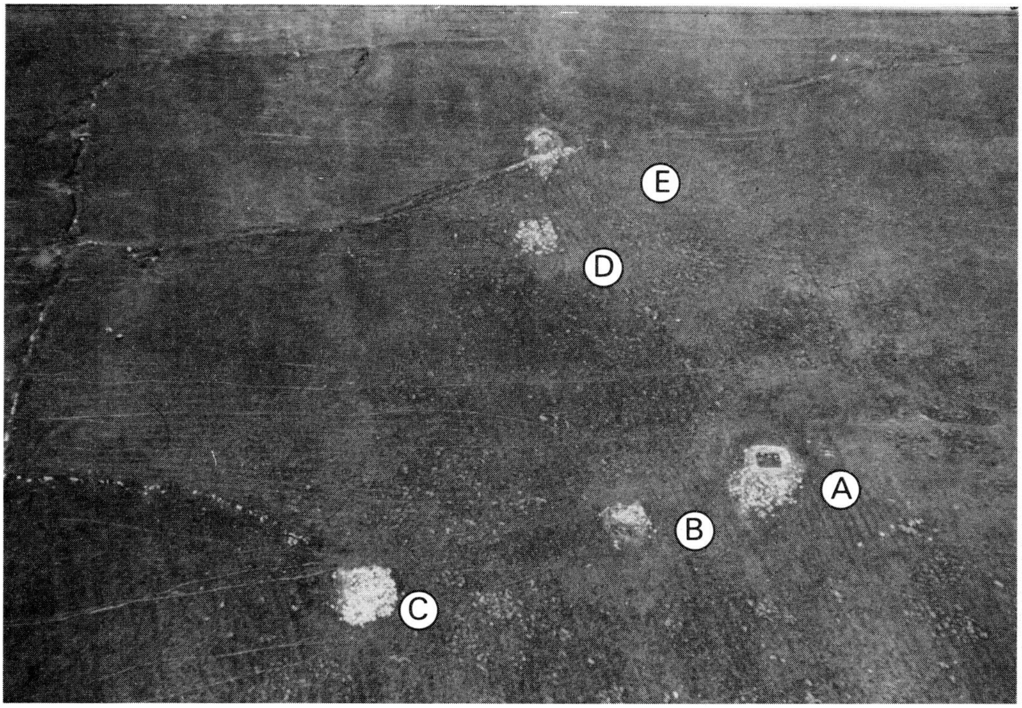

Bray South Slope, group of clochauns. 45
These drystone buildings have been described both by Westropp and by Françoise Henry. A and C are the best preserved; they were rectangular corbelled houses with the doorway in the east end; some stones in A carry incised crosses and others abstract designs. E, now very much damaged, was a round corbelled house; the wall that now runs diagonally through it is obviously younger. On the left this wall meets an older boundary fence.
Almost the whole area is covered by cultivation ridges which are younger than the buildings which interrupt them. North-west of C, above the curved wall, two sets of cultivation ridges intersect at right angles; the curved wall has a prominent lynchet on its upper side.

6°), and is built up of shaly material which has been moved downslope by frost action. The shaly debris is open in texture and well drained, and here peat did not form. At the cliff edge at the bottom of the slope rising water makes the ground wet, and here black bog-rush (*Schoenus nigricans*) flourishes. Although the soil here, as elsewhere in the island, is very poor in nutrients, the well-drained part of the lower slope, where the soil has some of the characters of a brown earth, was extensively cultivated in the past.

On the lower slope there is a group of five drystone buildings, 44
clochauns, which have been described by Westropp and by Françoise Henry. In two of them (Westropp's A and C) the masonry is as good
as in the corbelled oratories of the region, but as their doorways are in 45
their east ends they are not likely to have been churches. It has been
recently noted, however, that A has an incised Maltese cross on a stone 46
in the south wall of the entrance, and very lightly incised crosses (a and

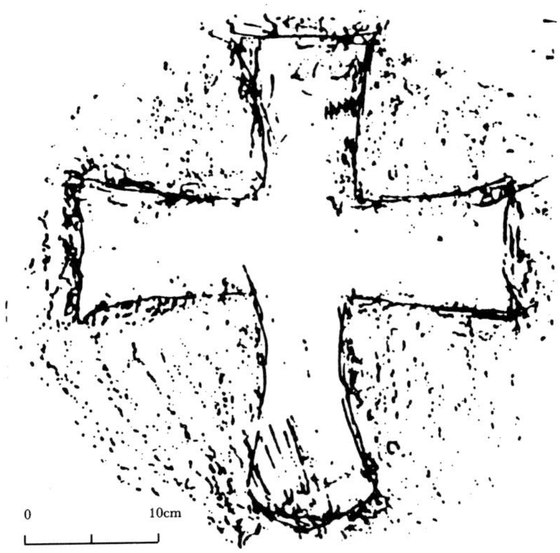

46 *Bray South Dip-slope, clochaun site. This rubbing shows a cross incised on a stone low down on the left-hand side of the entrance to Clochaun A.*

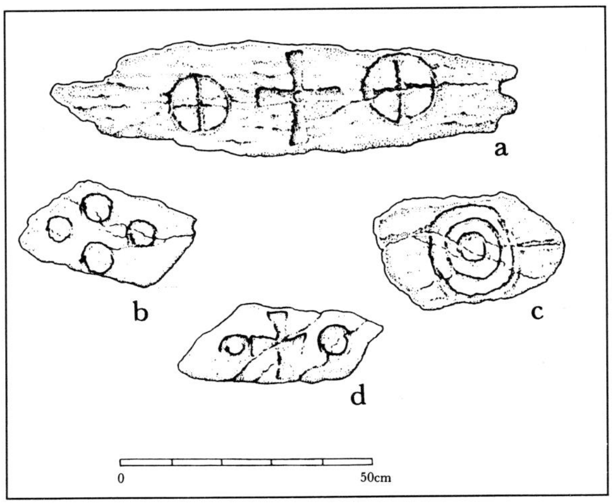

46 *Bray South Slope, clochauns. Small incised designs inside Clochaun A: a and d, crosses with circles; b and c, abstract designs.*

d) occur on some of the stones in the inner face of the north wall; Françoise Henry noted two geometrical designs (b and c). Similar crosses have been noted inside oratories in the Dingle Peninsula. Are these structures farmhouses? The style of masonry is similar to that of the buildings on the Cool East Lobe. There is a ruined corn-drying kiln further down the Slope; the rectangular arrangement of its passages has conceivable affinities with some Romano-British kiln types.

Although no forts now survive on the Slope we have three fort names along its coast. From east to west these are Doonroe Cliffs, Danganbeg, and Danganmore Head. Were there promontory forts here which have been carried away by the erosion of the cliffs they stood above? In south-west Kerry heavy cliff-falls are of frequent occurrence. In support of the idea that there has been heavy cliff erosion, we can note the
53 intermittent presence of a cliff-top trackway, known in local folklore as *Bóthar na Scairte,* the Route of the Cataclysm, a route which, according to the story, formerly extended along the whole length of the island.

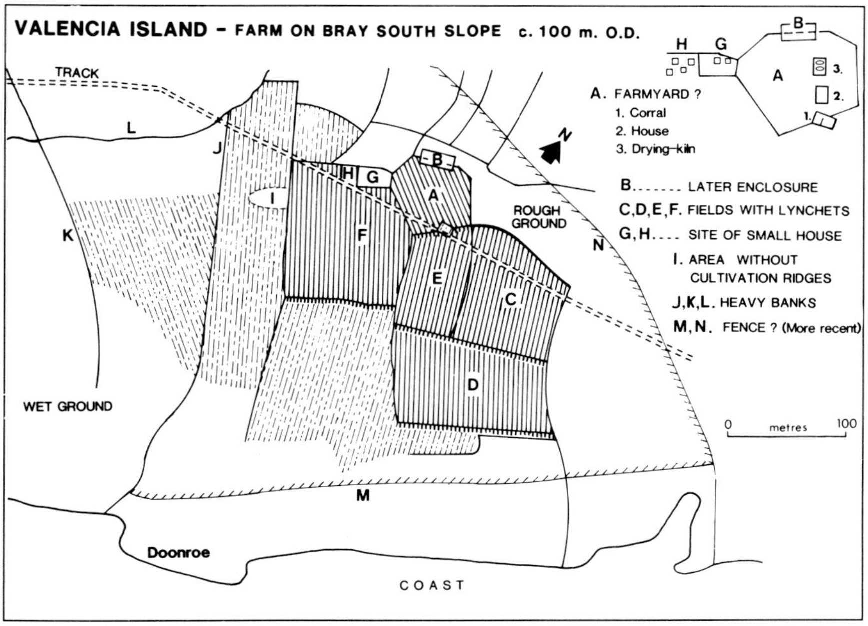

There are very curious features here, which I interpret as indicating the former presence of a plough-using farm. It is at 100m OD. 47

A is the farmyard, tucked on its northern side into a shelter-giving recess in the hillslope, and surrounded by a bank of earth and large stones. In it are (1) a rectangular corral, (2) the rectangular foundations of a house, and (3) the foundations of a 'keyhole' drying-kiln. B is a small enclosure at the back of the recess, built on top of the earlier bank of A. C, D, E and F are fields with high cultivation ridges and well-developed lynchets at their lower ends. West of these fields is an area with smaller cultivation ridges. G and H are small enclosures with the rectangular foundations of smaller houses. I is a small area where the cultivation ridges are absent; there may have been structures here. J and K are heavy banks of earth and stone; K may mark the western limit of the farm. L is also a heavy bank, perhaps the farm limit. M and N are marked as fences on the first OS map, but I do not think they relate to the farm.

What survives here is a winding track, about 1.25m wide, and guarded in some places by a stone kerb on the cliff side. If we follow it eastwards it starts in the embayment north of Danganmore Head, disappears at 53
the Head, but reappears on its east side, and then runs more or less continuously for a distance of about 500m, skirting the bay north-east of the rock called Danganbeg. It is then abruptly truncated by the cliff edge, owing to a rock-fall.

The drystone huts lie almost centrally on the Slope. To the east, just below the ridge-crest that separates the South Slope from the East Slope, there is a remarkable farm unit, of unknown age, cut through by the track that runs up to the signal tower on Bray Head; the tower was built 44
in the early eighteenth century; the track was in existence in 1810. 47

Tucked into a low, gently-sloping recess in the hillslope, at an altitude of about 100m and facing south-east, there is a polygonal enclosure (A),

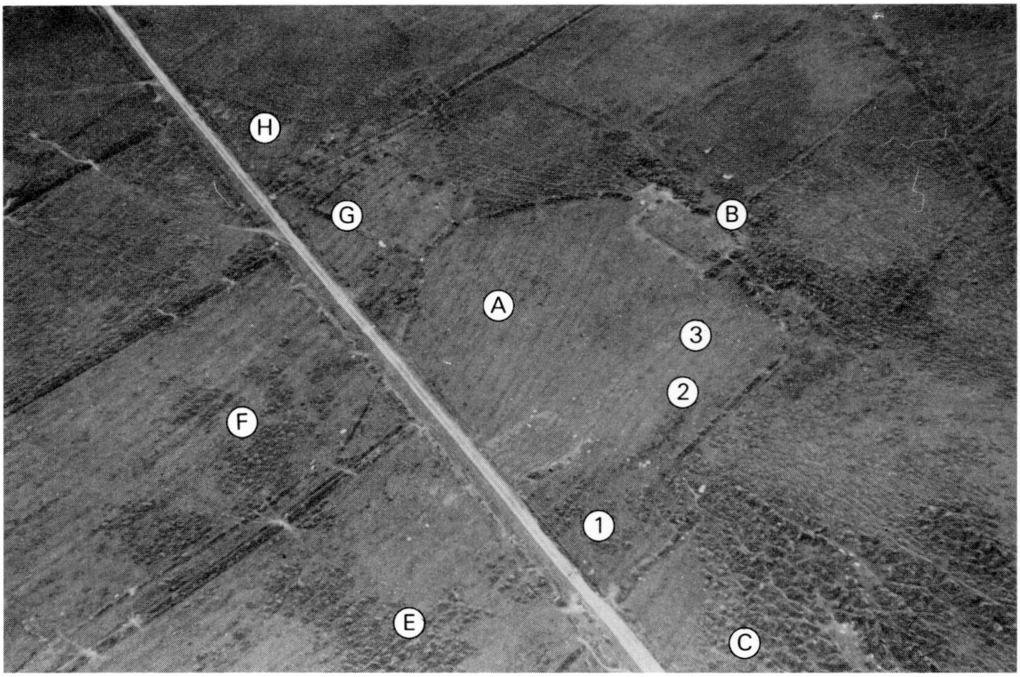

48 *Bray South Slope, farm (age uncertain).*
The track to the tower on Bray Head runs across the site; the track was probably built in the early eighteenth century. A is a polygonal farmyard, with inside it (1) a pound or corral, (2) a house site, and (3) the ruin of a corn-kiln. B is a later enclosure. C, E and F are fields with high lynchets at their lower ends. G is a small enclosure with house sites. H has house foundations beside it. The cultivation ridges in Field F clearly underlie the track, and must be older than it.

rather more than 100m across, surrounded by a heavy bank of earth and large stones, still about 1m high. It was a farmyard. In the middle of the south side, where the bank falls away in a marked lynchet, there is a rectangular enclosure (about 25m x 20m), surrounded by a similar heavy bank (1). The track to the signal tower grazed its south side, which was largely cut away when the track was constructed. It may have been a corral.

About 7m north of the north wall there is a rectangular (7m x 14m) house platform (2), with a suggestion of an entrance in the east wall. Some 5m north of the north-east corner of the platform there is a curious feature (3), about 8m long and 3m wide, with banks up to about 0.50m high; within are two conical hollows separated by a sinuous bank. I identify this feature as a corn-drying kiln of keyhole type. In a later modification on the north side of the main enclosure a rectangular enclosure (B) (30m x 7m) was superimposed on the original bank, which can still be traced across its floor.

From the west side of the main enclosure there springs to the south-

west another small (40m x 20m) embanked enclosure (G) with a marked 47
lynchet on its south-east side; in its north-east corner it has some puzzling
features — the outline of a small rectangular house, a curved wall, and 48
a small pit, possibly a 'borrow'-pit. Immediately to the west again, in the corner of a field (H), are the earth and stone foundations of four houses; two were substantial buildings (12m x 4m), divided into two rooms. Unfortunately the farmyard itself, the corral within it, and the adjoining enclosure have all been disturbed by cultivation ridges of a later date.

Downslope from the farmyard are four fields (C, D, E and F); all
have cultivation ridges with high lynchets at their lower ends. That there 53
had been extensive movement of soil downslope to build up lynchets was shown by the fact that at the upper end of Field F immediately below the track there was 0.35m of dark brown loamy, friable, free-draining soil resting on broken shale, while at the bottom end there was 0.80m of essentially similar material above the rock debris. That the soil here had not been disturbed in recent times was shown by renewed soil formation in the upper layers. Below 12mm of brown loam there was an incipient iron-pan 1mm thick resting on 8mm of loam, yellow-red in colour owing to iron enrichment. Also the cultivation ridges in the part of the field above the track continued without interruption below the track. I believe that the soil movement must have been brought about by the use of heavy ploughs.

If we try to define the farm area, Field C on its northern side ends against rough ground; on its east side we have a heavy bank, and this continues down the side of Field D and on to the coast. This is, I think, the eastern limit of the farm. Other fields with cultivation ridges but no lynchets wrap around the south-west sides of Fields D and F, ending either in wetter ground below or the steeper hillside above. On the west side of Field F there is a small area, I, which is not cut by cultivation ridges; there may once have been some structures here.

A heavy bank, J, faced with stone on its western side, then runs from the hillslope above down to the coast just west of the promontory named Doonroe. Above it a bank, L, runs at an angle to the slope, and so blocks surface water running down the slope. The water flowed against the north face of Bank L and the west face of Bank J, and in doing so cut a deep trench in which it now flows as a stream; its channel is the result of the construction of the bank. Further cultivation ridges lie west of Bank J. Rather less than 200m west of Bank J there is a further bank, K, of similar construction, with a stone-faced scarp to the west and thrown-up earth to the east; I believe this to be the western limit of the farm area, as no similar banks can be seen farther to the west. Here we are approaching the cluster of drystone buildings already described; a considerable area of cultivation ridges surrounds these buildings also.

Above the central farm unit, a number of banks run upslope over the crest of the ridge; there are no cultivation ridges between them. A fence, M, is shown on the 1847 six-inch OS map running up the crest of the ridge, but I do not think that it is associated with the farm unit.

A fence, N, is also shown parallel with the coast, but again I do not
47 think it is related to the farm unit.

48 I cannot suggest any age for the farm unit, beyond saying that extensive downslope movement of soil and the creation of marked lynchets would seem to imply the use of heavy ploughs, unknown in Ireland before the twelfth-century Anglo-Norman invasion. I did not see any mortar in any of the ruined structures. The cultivation ridges in the farmyard and in the corral within it must belong to a time after the farm had been abandoned.

The whole history of the South Slope is extremely puzzling. There was certainly intensive use in Early Christian times when the drystone buildings were in occupation, but were these the people who used the kiln further downslope? Do any of the cultivation ridges that now surround the buildings go back to this period, or are they younger? If younger, how much younger? Do they go with the farmyard to the east, or are they nineteenth-century? If they are late, where are the clachans of the cultivators? By island standards, the nearest clachan at Clynacartan was a small one, not at all proportionate to the great areas of cultivation ridges to the west. If the operators of the farm did have ploughs, how did knowledge of the plough vanish from the area? Excavation of the farmyard might solve these mysteries.

There are also other antiquities of uncertain age. Among the cultivation ridges of the farm unit are the rectangular stone foundations of two houses; the ridges swerve to avoid them. Just west of the farm unit in a patch of wet ground with a spring is the round foundation of a small hut, and there is a second similar foundation at the tip of Bray Head itself. West of the first hut there is a curious ruined slab structure. Two parallel lines of vertical thin slabs extending for some 6m and about 1m apart are linked by transverse slabs. There are also many walls, some running up onto Bray Top. It seems likely that this attractive slope would have been occupied in prehistoric times.

35 **19 Bray Top 15ha Bray**

61 Bray Top is a shallow dome, with its highest point about 275m. It is covered by blanket-bog. At one point the following section was recorded:

0-30cm	Brown fibrous peat, with modern roots of *Molinia* and other plants
30-120cm	Highly humified brown fibrous blanket-bog peat, with twigs of ling
120-140cm	As before, very highly humified
140-148cm	As before, but paler in colour, with sand and grit
148-155cm	Brown mud-like peat with some fibre and traces of sand
155-160cm	Mor — grey-black silt with small stones and high organic content; near top thin black horizon, about 1mm thick, presumably rich in charcoal debris; becoming paler in colour downwards
At 160cm	Surface of dark brown sandy loam with stones

The base of the mor had a ^{14}C date of 4790 BP. A short pollen diagram showed that when the base of the mor and the soil immediately below it were forming the area was covered by open vegetation, part heath and part weedy pasture, grazed by Neolithic herds.

Between 140cm and 148cm the peat was paler in colour, with a content of sand and fine grit. This foreign material must have been carried in by water or wind. In view of the fineness of the material, it was probably transported by wind from bare ground, possibly man-made, nearby.

Some old walls run north/south across the Top, ending at the lip of the Bray North Slope. One wall is largely buried by peat.

There has been some sporadic peat-cutting, possibly associated with the signal tower and other buildings at the tip of the Head.

20 Bray Mega-cliffs Bray

This massive range of cliffs, which rises to a height of 275m, truncates the western end of Bray Head.

21 Bray North Scarp-slope 20ha Bray *35*

This is a steep (about 30°), straight, block-strewn scarp-slope facing *61*
north; it is the north face of the presumed Bray Horst. It runs down into the extreme western limit of the Imlagh Basin, and ends to the west at the cliff edge at Foilanean.

Some stone walls run down the slope and out under the bog of the Basin; there are also some animal shelters tucked in among the blocks.

THE NORTH RIM OF THE IMLAGH BASIN

22 West Rock Rim 5ha Bray *61*

Before moving into the Basin itself we complete its perimeter, moving along the coast to the north-east.

This narrow rim of peat-free ground starts at the cliff edge at Foilanean, and runs past Beenaniller Head — the last home of the sea-eagle in Ireland — and on to Coosagauna; it slopes eastwards.

Near the centre of the Rim a short (5m) line of boulders suggests the ruins of some type of structure. The tip of the Head is cut off by two short ditches; one is probably modern.

Two rectangular areas with cultivation ridges can be clearly seen in aerial photographs; I failed to trace any corresponding features on the ground.

23 Coastal Downland 30ha Bray, Coarha Beg *61*

As we move north-east past the inlet, Coosagauna, glacial gravels appear on top of the rock and continue as far as Coosheennacushduff. In places
they slope gently southwards from the cliff edge; elsewhere they rise into *71*
a ridge, a little inland from the coast. In each case they provide the north rim of the Basin. At one point near the cliff edge there was a mound,

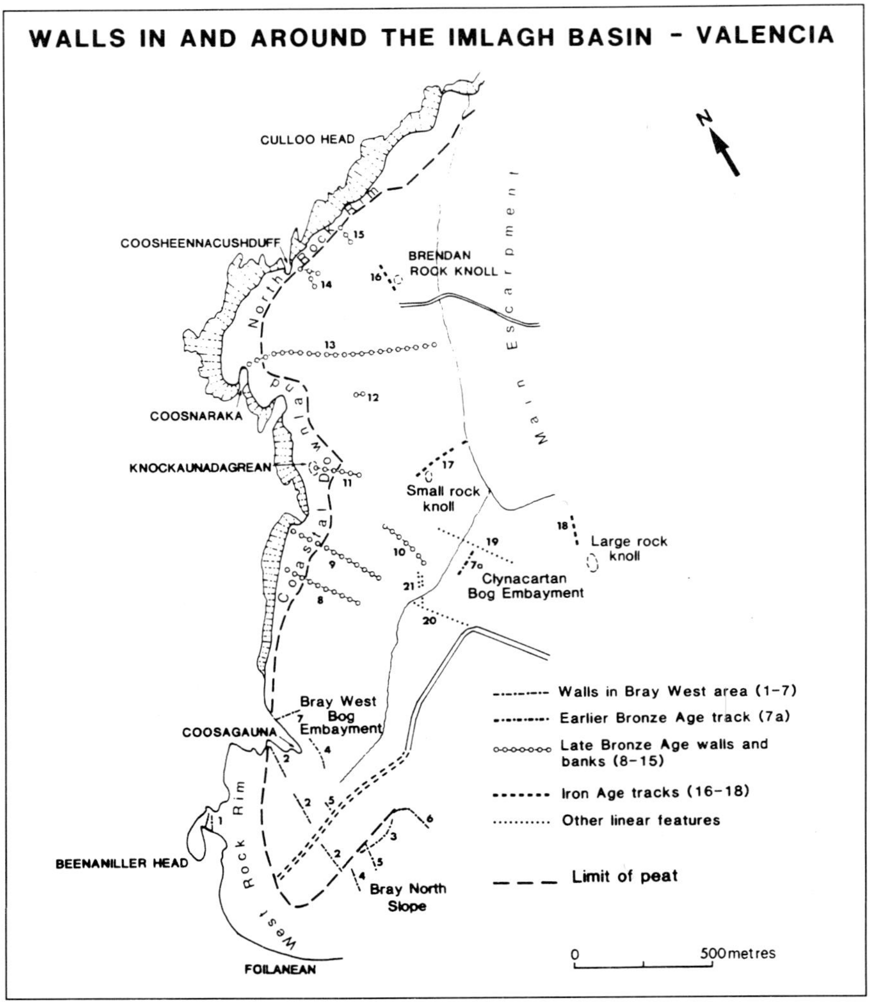

52 *Coastal Downland and seaward side of Imlagh Basin.*
Plan of walls and tracks partly buried by peat. There seem to be three groups, one in the Bray West Bog Embayment, one on the Coastal Downland, and one near the main drainage stream of the Basin.

West Embayment: with the exception of Wall 1 near Beenaniller Head, the other walls (2-7) appear to divide up the Bray North Slope and the West Embayment.

Coastal Downland: walls 8, 9, 10, 11, 13 and 14 run south-east down the Downland towards or into the peat of the Basin; 15 is perhaps an outlier. These walls are sub-parallel and include large stones; they appear to be Late Bronze Age, as where 13 ended in the bog it had a ^{14}C date of 2680 BP. South-west of 13, and linked to it by a connecting wall, was an area of stone-walled fields, 12; this area had a similar ^{14}C date.

Near main drainage stream: near a small rock-knoll west of the stream, peat underneath a track of earth and stone (17) had a ^{14}C date of 2020 BP. Near a large rock-knoll south of the stream a track of similar material (18) had a ^{14}C date of 1700 BP. Near the Brendan Rock-knoll a stone trackway in the bog (16) had a ^{14}C date of 1720 BP. All these dates are Iron Age, and indicate occupation of the area around the bog at that time. A bank (19) built of sods is perhaps relatively recent; 20 and 21 are enigmatic.

A stone track (7a) had a ^{14}C date of 2920 BP and is Earlier Bronze Age.

Bray South Slope, Bóthar na Scairte. 53
This ancient cliff-top path, protected by vertical slabs on its seaward side, is seen running westwards towards Danganmore Head; to the east it is cut off by a cliff-fall. It may be part of a legendary road that ran from the mainland side of Doulus Bay across Beginish and Valencia to the Skelligs; according to the story the lower parts of the road were swept away by a tidal wave, hence the name Scairte, *cataclysm or disaster. Games were in progress when the wave struck, and a point of land in Doulus Bay is still called* Rinnanenagh, *the promontory of the* aoneach *or festival.*

Bray South Slope, farm. 53
In the foreground is the high lynchet at the lower end of Field F of the farm. The track to the signal tower on Bray Head runs up across the top of the picture.

54 *The north slope.*
Following the coast from the bottom, we see the North Rock Rise with its structures, then (pointing away and foreshortened) Dundagallaun, then the bank and ditch of Doonroe, and beyond the Cool and Tinnies Slopes rising into Geokaun Top. The Fogher Cliffs, which show a recent rockfall, cut into the Top. One-third up the right margin, we see the scarped end of the Cool West Lobe, and two-thirds up we see fields on the Cool East Lobe where a group of clochauns lie. The mountains of Kerry fill the background.

now largely quarried away, of finer glacial gravel, Knockaunadagrean. 71
It gets this name, the Hill of the Two Suns, because in summer both the rising and the setting sun shine on it; *púcaí* (fairies) are reputed to have played football and run races there.

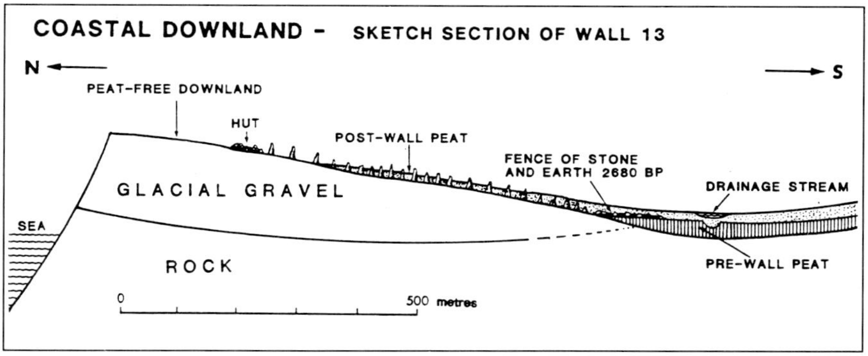

Coastal Downland, Wall 13. 55
Sketch-section of the most impressive wall, 13, which rests at its southern end on already-formed peat in the Basin, and is buried by later peat. This end of the wall was dated to 2680 BP.

Coastal Downland, Late Bronze Age wall. 55
This wall of stones (No. 13), some of which are quite large, runs down the slope of the Coastal Downland and ends buried in peat in the Imlagh Basin. In this stretch it was formerly buried by peat, and the second group of stones in the wall is still partly buried by uncut peat. Where it ends in the bog the contemporary layer of peat had a ^{14}C date of 2680 BP, which places the wall late in the Bronze Age.

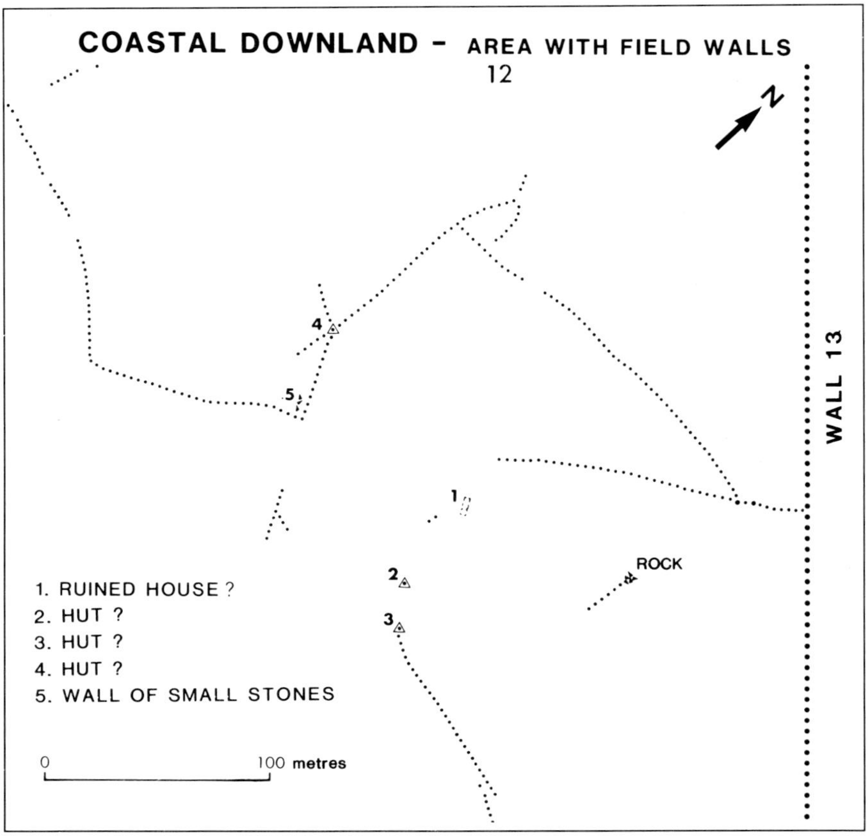

56 *One of the walls in this area connected with 13. The area, which had a radius of about 200m, had several walls which appear to have bounded polygonal fields. There were also what appeared to be some ruined structures; 1, which was still partly buried by peat, may have been a house; 2, 3, and 4 may have been huts; there was a suggestion of a corral immediately south of 4. At 5 some walling of small stones was still buried by peat.*

One would think that this area, where the soil rests on gravel, would be well-drained, but a profile near the cliff edge showed 15cm of peat, and then 30cm of poorly-permeable leached silt resting on stones. The soil is a peaty gleyed podzol. Wind and sea-spray have probably prevented the formation of a peat-forming plant community near the cliffs, and we thus have here a coastal fringe of a fine sward — a kind of downland — not of grass but of dwarf plantain, a type of vegetation well developed on Ireland's western coasts, called *Plantago*-sward by Praeger. I call this area the Coastal Downland.

This sward will have provided grazing at all times, and there are several linear stone walls which run south from the cliff edge across the downland under or onto the peat in the Basin. In earlier times, before the expansion of peat growth, the Downland was considerably larger than it is today, and its soil was less degraded, probably carrying grass, after its primary woodland had been removed by man. Where the walls enter the bog,

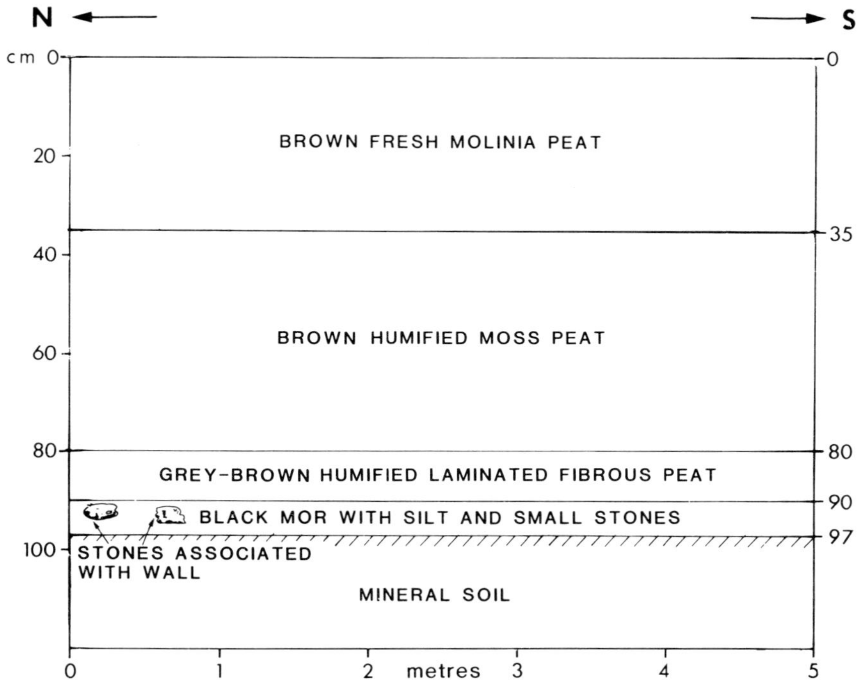

Peat face, Area 12. 57
This sketch-section of a peat face at Point 5 in Area 12 shows at the base a layer of mor with wall stones embedded in it. Samples below and above the stone had ^{14}C dates of 2900 and 2050 BP respectively. The basal peat was highly humified, and humification decreased upwards. The surface layer was blanket-bog peat. There was no fen-peat.

large stones tend to be replaced by earth and smaller stones, and the walls become banks. In the bog they usually rest on peat, and have been buried by the further growth of peat.

One wall and bank (No. 13) is much the most impressive of the walls which run across the Downland. It is essentially a single line of large stones, some of them upright slabs about 1m high, set at intervals 55
transverse to the line of the wall; smaller stones fill the intervening spaces. On the Downland the wall runs for 650m, and it continues out into the bog for a further 100m. It begins on the coast at the inner end of Coosnaraka and first runs south-east for 100m, but it has been robbed of its stones in this section and is represented by sockets only. It meets the ruins of a circular, roughly-corbelled stone hut, 3m in diameter, and turns south-south-east, running down the gentle slope of the Downland for a distance of about 550m. At the time of its erection marsh-peat had already formed on both sides of the drainage stream in the Basin.

When the wall reached the edge of the bog it changed into a bank as it continued for about 100m out onto the bog, small rounded boulders and earth being used in its construction. Some sand and small stones spread out on the adjacent bog surface. Peat growth continued, and as it thickened it buried the end of the bank and also the outspread sand; the sand layer had a ^{14}C date of 2680 BP, which places the bank and wall at the beginning of the Dowris Phase of the Late Bronze Age. As the peat thickened further, its edge crept northward up the Downland, progressively engulfing the stone part of the wall.

Immediately west of the southern end of Wall 13 there is an area with
56 field walls (No. 12) which links up with 13, and is presumably of the same age. There are a number of fences, mainly of large stones but with some sections of earth and smaller stones; in places they rest on mor.
57 At Point 5 a small basal wall stone was embedded in mor and buried by peat. The mor immediately below the stone had a ^{14}C date of 2900 BP, and the base of the peat resting on the stone had a ^{14}C date of 2050 BP, which is compatible with a Late Bronze Age date for the emplacement of the stone and thus the construction of the wall of which it forms a part.

If intact, the walls would enclose about five polygonal fields with a total area of about 1.25ha. There are four ruined stone structures (1-4), one of which (3), still largely buried by peat, may be the remains of a house; a small enclosure or corral (beside 4) is also suggested. Here we may have a Late Bronze Age farm.

52 Altogether six linear walls or banks, spaced 200-500m apart, without cross-connections, run south from the cliffs across the Downland towards the Basin. Number 8 is a well-defined length (400m) of stone wall which runs down into bog; No. 9 is a long length of wall alternating with ditch and bank from the cliff edge south almost to the tributary stream from the west. Fence 10 could not be traced near the coast, but started as a ditch and bank about 300m inland, and continued for about 300m almost to the stream; in places the ditch and bank were cut in peat, which had mor at its base. Number 11 is a 300m length of stone wall which starts from Knockaunadagrean and runs south into bog. Wall 13 (already described) is 500m east of Wall 11, and the field system (No. 12) lies between them; east of Wall 13 is a tangle of slabs (10m x 6m), perhaps the last vestiges of a house. Number 14 is a short (150m) length of wall running in from Coosheennacushduff; the wall is double in places, and passes a cluster of slabs largely buried in peat; other short lengths of wall are nearby. The wall ends on *Phragmites*-peat, and there is no mor. Number 15 is a short (100m) length of irregular wall which starts from a badly-defined stone platform close to the coast; mor is present at the north end of the wall, but to the south the peat quickly thickens, with *Phragmites*-peat at the base.

This series of walls divides up the agriculturally valuable Coastal Downland, and if we accept the ^{14}C dates then these features must belong to the Dowris Phase of the Late Bronze Age. We have one archaeological find which shows that Dowris folk were on the island.
99 Towards the end of the nineteenth century a Late Bronze Age sword

North Rock Rim, corral. 59
On a narrow band of smoother ground between rock outcrop (above, right) and bog (below, left) we see a flimsy circle of stones, about 40m in diameter. It can hardly have had a defensive purpose, and may be a corral for animals. If we regard the circle as a clock-face, at XII we see a modern sheepfold, at VI a round hut linked to the circle, and at IX a complex of small buildings.

was found during peat-cutting in the Basin (in Coarha Beg Td). The pollen diagram from the Brendan Rock-knoll nearby shows some features 15
of vegetational disturbance that might be ascribed to farming at this time.

24 North Rock Rim 15ha Coarha Beg, More 52
East of Coosheennacushduff the gravels end, and rock sloping southwards 61
from cliffs about 30m high, which continue as far as Shrone Point, forms the North Rock Rim of the Basin, essentially a dip-slope. The stream that drains the Basin cuts through the Rim just east of Culloo Head. At its exit it breaks up into small channels across a shelf of rock, and drops into the sea over cliffs about 10m high. Beyond Shrone Point the Basin lacks a boundary rim for a distance of about 300m, and the slope to the north continues to the cliff edge. The ground then rises again, and the Cool East Lobe restores the rim.

Immediately south-south-east of the Triangulation Point 101 near Culloo Head a lightly-built stone wall surrounds a circular area about 30m in diameter; some hut-like features are included in the wall, which is probably a corral of some type, possibly Early Christian in age. Of 59
unknown age are some remnants of stone walls which run south down the slope of the Rim underneath the peat of the Basin.

Just east of Shrone Point there is a deep narrow gully, which formerly had a flight of rock-cut steps leading down to the sea below, but the

steps have been destroyed by erosion; fragments of a few can still be seen. Off the rocks here fishing is still very good, and in former days there was a fleet of small boats based here. The rough track which leads
70 down from the road at Coarhabeg to Shrone Point was constructed for the benefit of the fishermen.

THE IMLAGH BASIN

60 **25 Imlagh Basin 515ha Coarha Beg, More, Cool West**
Today at first sight the Imlagh Basin is a peat-filled basin surrounded
61 by inwardly-dipping slopes. But in pre-peat times we would have seen a shallow arcuate stream winding its way round a rock ridge or
4 escarpment — the Main Escarpment — in the centre of the Basin. This escarpment is, as we have seen, a salient from the Cool West Lobe which projects into the Basin. As a result the course of the stream from source to sea is twice as long as the crow-flight between the two points.

The Basin also held a number of other rock rises, smaller in scale, but these did not affect the drainage pattern. These rock rises provided

60 *Imlagh Basin, general view.*
Here we see the whole sweep of the Basin, with Beenaniller Head in the bottom right-hand corner. In the foreground on the left we see the peat-free Coastal Downland sloping down below the much-cutaway peat of the Basin. At the point where the left margin of the picture is trimmed, the Downland gives way to the North Rock Rim, also peat-free. A block of fields is seen in the right centre; these are on the Main Escarpment where the peat has been cut away. Beyond the Basin we see Geokaun Top, with the lower Kilbeg Top to the right. The mountains of Kerry fill the background.

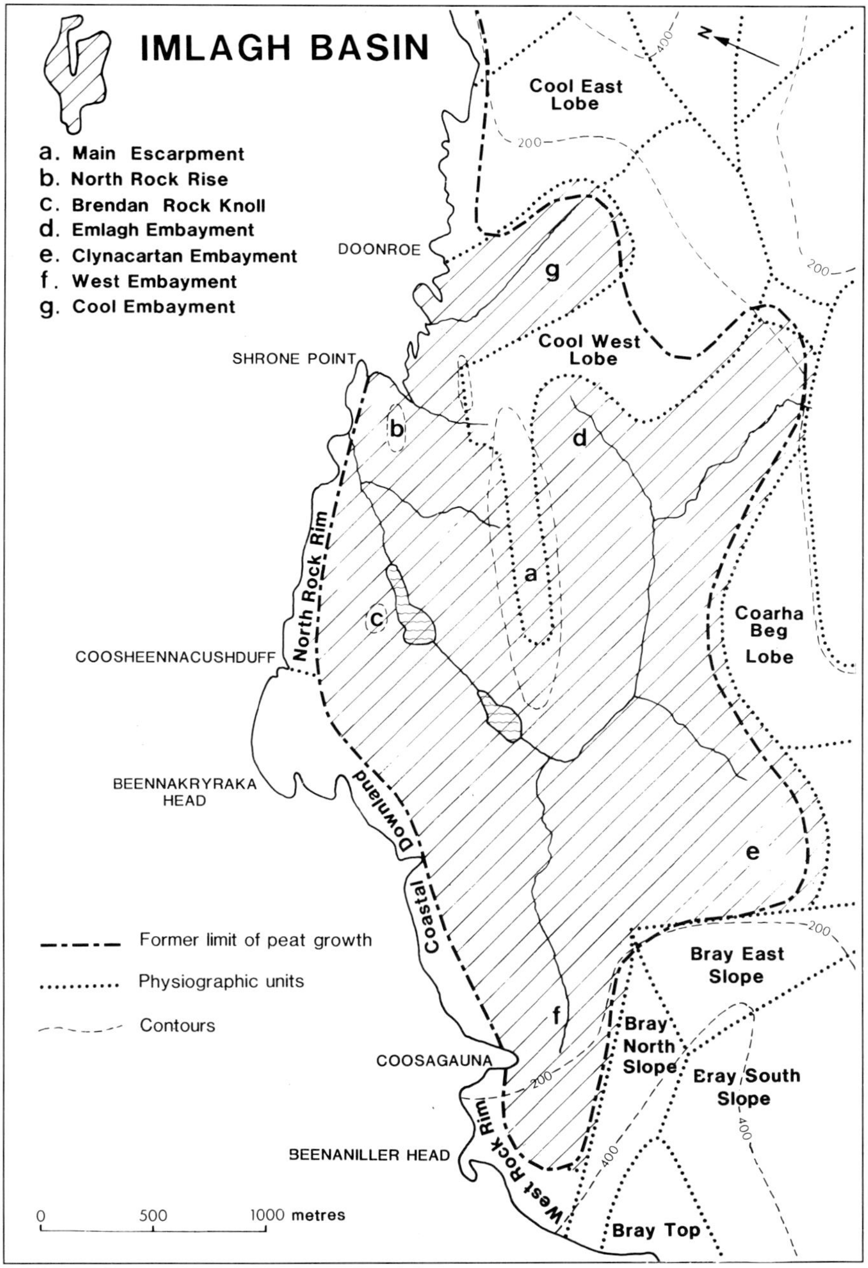

Map of Imlagh Basin.

62 *Imlagh Basin, Clynacartan Bog Embayment.*
Peat beside the stone slab projecting from the face of a bog drain had a ^{14}C date of 2920 BP, so it must have been emplaced in the Earlier Bronze Age. A second slab, removed in drain-cutting, lies on the bog surface. They perhaps formed a track (No. 7a) across the bog.

Behind the top of the ranging-rod the drain cuts into the side of a track (No. 19) or bank built up by laying sods on the bog surface; the layers of sods can be seen in the picture. The track is 1m wide and 25cm thick and is buried by 30cm of peat; it runs in a straight line for 2km. It is probably relatively recent in age.

61 attractive sites for early man. I regard the rises as resulting from differential sub-aerial erosion during the Tertiary.

The stream took its origin in springs at about 70m in the low ground north of Coarha Top. It started off in a west-north-west direction, but was forced to turn west-south-west as it ran up against the east end of the Main Escarpment — the beak of the duck-shaped Cool West Lobe

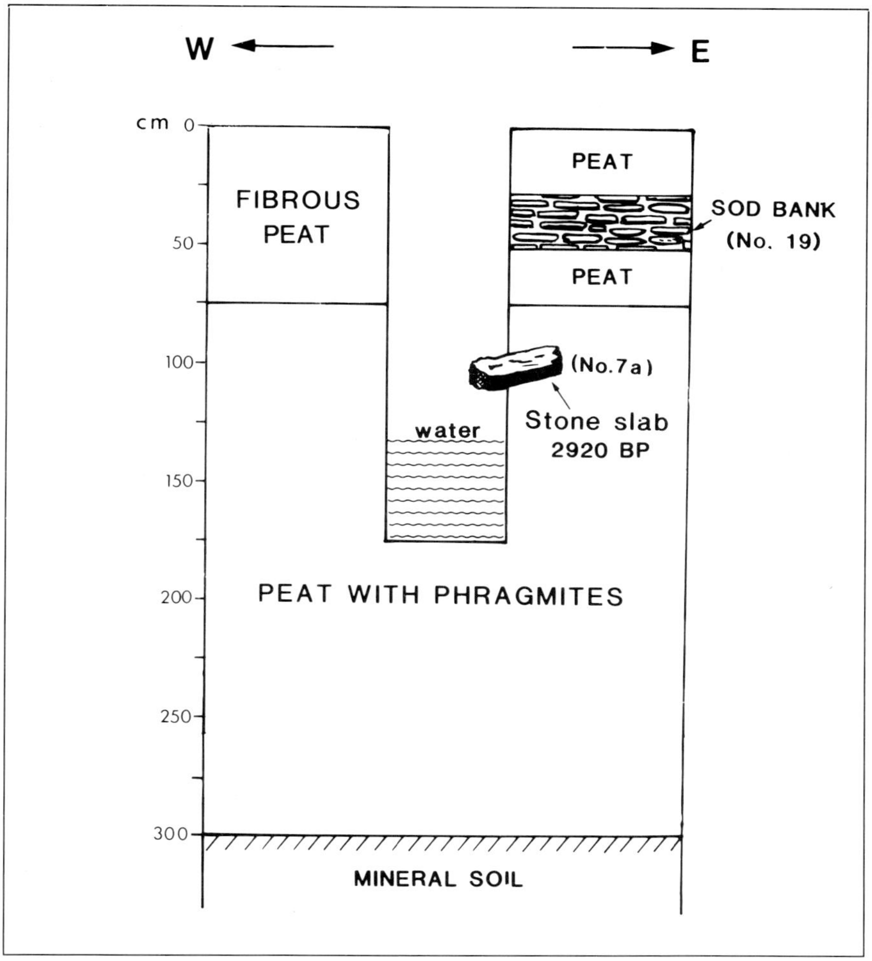

Imlagh Basin, Clynacartan Bog Embayment. 63
Sketch-section of bog drain showing stone trackway of the Earlier Bronze Age (2920 BP). Apparently a line of elongated slabs was laid on what was then the bog surface; a few stone fragments were embedded in the peat at the same level nearby. The same drain also intersected a bank of sods (No. 19) at a higher level. The base of the peat is perhaps Neolithic in age.

already described. As it turned it received a tributary which arose from springs in the area pictured as the low-lying cheek of the duck, the area I have named Emlagh. The stream then made a hairpin bend round the west end of the Main Escarpment. As it started to turn it received a tributary from the Clynacartan area, and as it made its full turn to the north-north-east it received another from the low ground north of the Bray North Slope. It then entered an area which was still very poorly drained when the first Ordnance Survey was made in 1842. Leaving this it flowed north-east, and collected a tributary from the north slope

61 of the Main Escarpment before reaching the sea near Shrone Point.

For a distance of 300m beyond Shrone Point the Basin loses its rim, and the ground slopes north to the cliff edge with a thin covering of peat. Here there are two small streams, one of which falls into the sea down a low waterfall at Coosatorinth, and the second in a similar fashion at Coosheenadagallaun. The second lies between the Cool East and the Cool West Lobes.

The courses of the main stream and its tributaries were later filled by peat which expanded into broad embayments, burying many archaeological features as it did so.

52 Between Knockaunadagrean and the main drainage stream, near to
the stream, a small rock-knoll rises through the peat. An earthen track (17) buried in peat, and apparently running towards the south-west end of the Main Escarpment, skirts the knoll. There was about 50cm of fibrous peat above the bank. There was then about 30cm of sand and small stones mixed with peat (the bank); near the base was a thin layer with brecciated lumps of firmer peat. Underneath was about 40cm of sand-free *Phragmites*-peat which rested on the mineral soil. A ^{14}C sample, taken immediately below the sandy layer, had an age of 2020 years.

About 500m south of the same end of the Main Escarpment a large rock-knoll (probably a relict inselberg) rises through the bog. A short distance north-west of the knoll a narrow earthen track, 60cm across
and 35cm high (No. 18), is being revealed by continuing peat-cutting.
71 From what can be seen at present it appears to be running north towards
the Main Escarpment. The track lies on 25cm of *Phragmites*-peat; the peat immediately above the track still has some content of *Phragmites*, but the uppermost surviving peat is fibrous and probably of ombrogenous origin. As spread from the track thinned out laterally, structures which may have been created by spades could be seen. A ^{14}C sample taken
where the spread had thinned to a washed-out sheet of sand had a date
65 of 1700 BP. It is not impossible that these two structures were in fact
one, a narrow track of sand and stones of Iron Age date running across the Basin from the large rock on the south towards Knockaunadagrean on the north, via the Main Escarpment and a minor rock-knoll.

About 500m to the north-west of the last site, near the tributary stream
65 from the Bray West Bog Embayment, a peat section showed two
superimposed features. The profile was:

0-30cm	Brown *Molinia*-peat
30-55cm	Bank 19, 100cm wide, built up of horizontal grass sods, with adhering sand and small stones
55-75cm	Brown fibrous peat
75-100cm	*Phragmites*-peat, rich in fibre
100-140cm	Large stone slab in a trackway (No. 7a); peat at the same level had a ^{14}C date of 2920 BP
140-270cm	*Phragmites*-peat, rich in fibre
270-300cm	Muddy peat, with less fibre; traces of sand at base
At 300cm	Top of mineral soil

The sod bank was No. 19 in a series of walls and banks already noted. It may be relatively modern, although it has in some places a thin covering of peat. It begins at Knockaunadagrean, and can be traced in a south-south-easterly direction for 2km.

The large stone slab at a lower level was probably one of a line of slabs laid on the bog surface as a trackway about 3000 years ago in the 62
Earlier Bronze Age. Two other slabs, removed when the drain was cut, 63
lay nearby.

Pollen counts in the basal layers of the peat showed grassland, with only a limited number of trees in the vicinity. Pollen of alder was not 16
noted, so the base of the peat should be more than 5000 years old, and the grassland may be Neolithic in age.

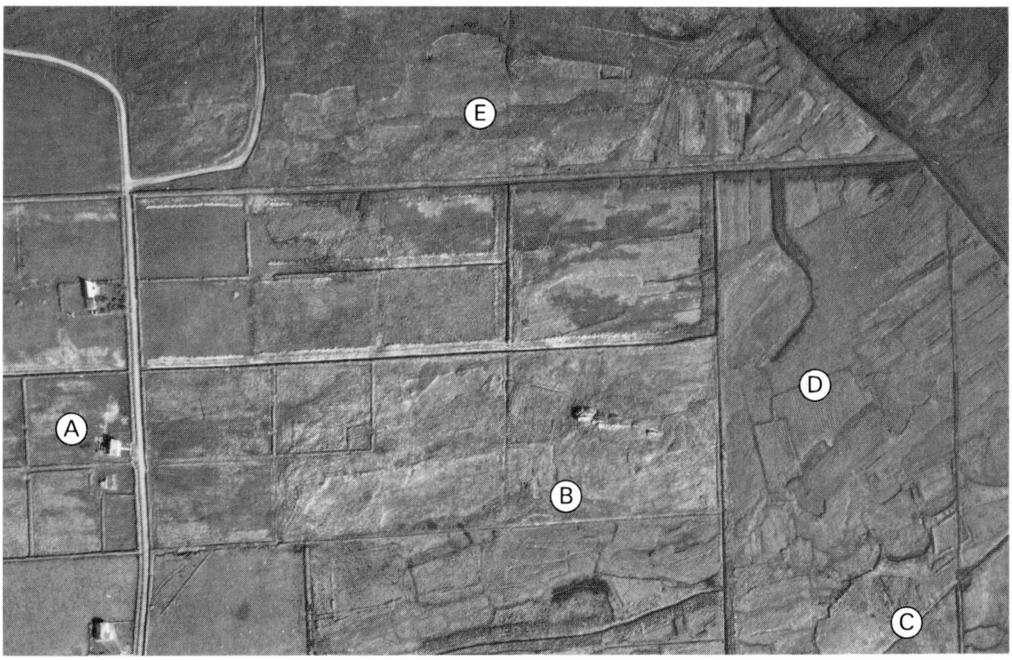

Imlagh Basin and Clynacartan Bog Embayment. 65
On the west (A) we see the road and settlement laid out on the east slope of Bray Head in the late nineteenth century. In the centre (B) we see a rock inselberg rising through the bog. In the south-east corner we see the arcuate bog cutting (C) in which a Mesolithic stone platform is exposed; immediately east of the cutting we see on the bog surface a diamond-shaped embanked 'garden' with cultivation ridges; a later drain cuts through the 'garden'. At D we see a bog cutting at the south-west end of which lies an earth and stone trackway dated to 1700 BP; south of the inserted letter cultivation ridges can be seen on the bog surface. At E a drain linking two bog cuttings showed a slab trackway (No. 7a) dated to 2920 BP; at a higher level the drain also cut through a bank built up of sods (No. 19).
Based on Ordnance Survey by permission of the government (permit 5149).

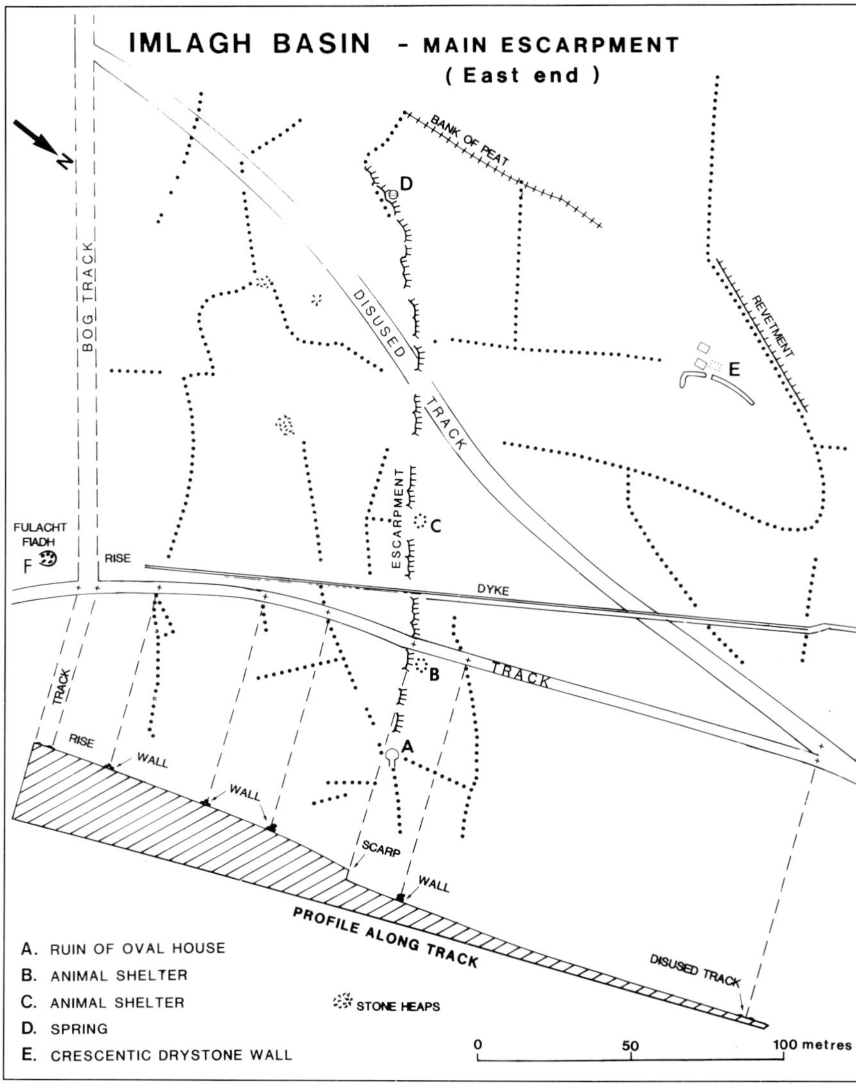

66 *In the limited area covered by the plan the highest ground is towards the left. The Early Bronze Age* fulacht fiadh *(F) with a* ^{14}C *date of 3250 BP lies here. Dropping downslope to the right we come to the first escarpment (about 3m high) running north-east/south-west. Along its line, A is the ruin of an oval house with annexe tucked against the scarp. In its vicinity are old walls outlining small rectangular fields. B and C are animal shelters similarly placed. D is a spring, and immediately beyond it the top of the scarp has been crudely revetted. Below the scarp the ground is very much wetter, but walls of large irregular fields can be traced through it. At E there is a crescentic drystone wall, probably of no great antiquity. Immediately west of it are some small house and hut foundations. North of it there is a low revetment about 75m long. Further downslope (not on plan) there is a second low rock scarp. Over the whole area covered by the plan there are lines of large stones which seem to enclose large irregular fields. There are also stone heaps, either the last remains of huts or stone piles from field clearance.*

The small creek, Coosheenadagallaun, on the north coast of the Basin, has already been referred to. It takes its name from a small promontory fort, *Dundagallaun,* on its western side, illustrated by Westropp. Here 54
a short rock-cut fosse and a stone wall cut off a slender peninsula about 15m long and 6m across. Access is by a central gangway flanked by two standing stones, each almost 2m high. There are no features within. As there are no signs that the peninsula has been reduced in size by erosion, and it is only separated from the mainland by a narrow cleft, not more than 10m wide, it is difficult to see what strategic purpose the fort might have served.

25a Imlagh Basin, Main Escarpment 15ha

Although unimposing in appearance today, this narrow rock rise, about 66
1km long, plays an important role in the drainage of the Basin. It is the south-west tip of the Cool West Lobe, and is aligned north-east/south-west parallel with the general strike of the rock beds in the area. The beds are dipping to the south at a small angle, and are cut across by low escarpments on the north face of the rise. As the Basin started to fill with peat the lower-lying part of the rock ridge that linked the Main Escarpment to the Cool West Lobe became covered first, leaving the Escarpment a relatively well-drained island, attractive to human settlement, especially on the eastern end. Continued growth of peat then buried it completely, and it has since been partly re-exposed by turf-cutting and drainage.

On the south slope a fine-grained muddy peat formed first, and then peat with *Phragmites.* Brown humified peat, perhaps ombrogenous in origin, followed. After peat had formed to a thickness of 200cm, a typical stone *fulacht fiadh,* with slab-lined trough and fireplace of smaller stones, 68
was erected on the surface of the bog. Burnt stone and charcoal 69
accumulated round about to a thickness of 25cm. There were flecks of vivianite throughout the debris, which indicates that bone was also discarded. The charcoal was of stems of ling only, which suggests that there was no large firewood available in the area; it was dated to 3250 BP, an Earlier Bronze Age date. Further peat growth then swallowed up and preserved the structure. Recent peat-cutting has now exhumed it. As the cuttings went deeper and approached the stony debris around the fireplace, the diggers abandoned their cutting. As a result the structure stands up on a pedestal of uncut peat surrounded by cutaway bog. It deserves protection as a national monument.

Just east of the point where the track to Shrone Point crosses the first step-like escarpment the ruins of an oval house (A) of well-built stone, 4m x 3m inside, with a small annexe to the east, were tucked in against the scarp. The house was surrounded by small subrectangular fields, and there were two animal shelters (B, C) nearby, also built into the scarp. Here we may see the layout of a small farm of Early Christian age.

West of the track were many further stone walls enclosing much larger irregular fields, with some piles of field stones or hut ruins. Where the

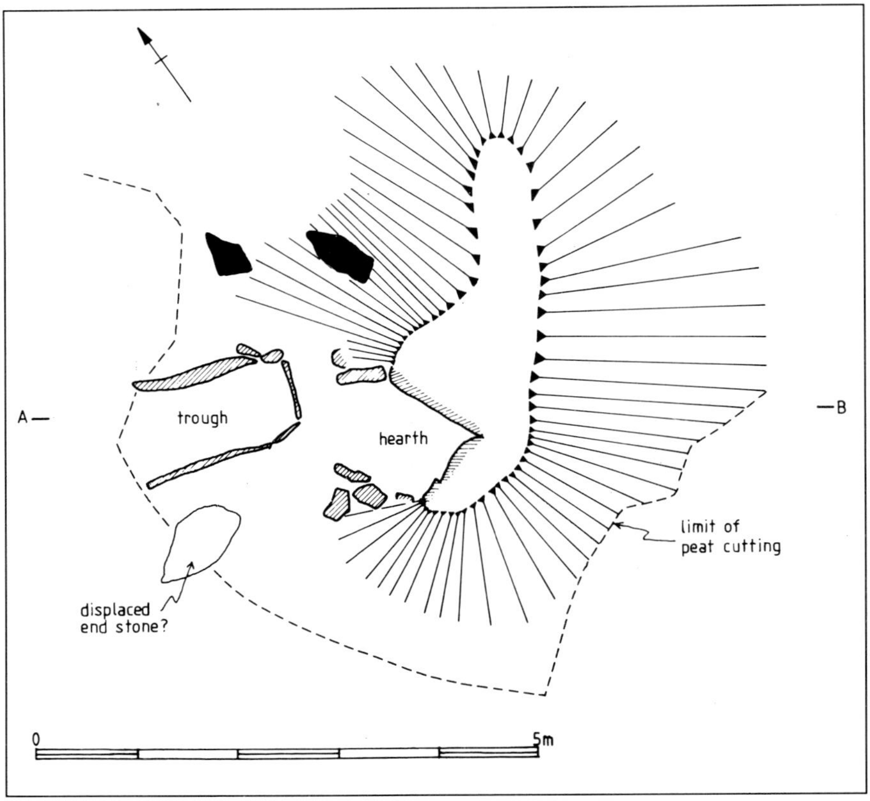

68 *Imlagh Basin, Main Escarpment.*
Sketch-plan of Earlier Bronze Age fulacht fiadh *(3250 BP). The end slab of the trough was missing. The two stones above the trough are probably part of a later wall system.*

escarpment bent around to the south there was a spring (D) at its foot; a little further on the top of the escarpment had been revetted with large stones. Below the first escarpment on the west of the track there was a flat wet area traversed by stone walls. This ended to the north at what appeared at first sight to be a rock escarpment, but closer inspection showed that it was a revetment of heavy stones. On the flat was a curious low, crescentic, roughly-built drystone wall (E); it was about 50m long and 1.5m high; it is probably not of any great antiquity. Immediately to the west of a gap in this wall were the ruins of a round hut and the earth-based foundations of three rectangular houses, 4m x 3m internally. This type of house I regarded as a late 'cabin', perhaps nineteenth-century in age.

There is a second, lower escarpment (not illustrated) a little further to the north.

At the south-west end of the Escarpment, west of the two abandoned cottages, there is a low round mound of stones, 10m in diameter, possibly a ruined cairn.

Imlagh Basin, Main Escarpment, fulacht fiadh. 69
We look along the trough of the fulacht fiadh *to the fireplace behind. The* ^{14}C *date is 3250 BP, placing the structure in the Earlier Bronze Age. The erect stone on the left probably belongs to a later wall. In the background we see the Coarha Top rising above the floor of the Basin.*

25b North Rock Rise 2ha Coarha More

Immediately south-west of Coosatorinth there is a low rise of rock in 54
shallow terraces facing north, with a thin interrupted covering of peat. Oval in outline, and about 200m long, it is aligned north-east/south-west. I call this the North Rock Rise.

There were numerous signs of human activity here, none firmly datable. Perhaps belonging to the Early Christian period were the ruins
of two two-roomed stone houses of poor masonry (A, G); one of them 70
was built against a rock overhang. There were also a ruined circular hut (E), a small robbed cairn (D), numerous lengths of stone wall, and some piles of stones.

25c Brendan Rock-knoll 1ha Coarha Beg

Near the northern margin of the basin, north of one of the areas previously noted as swampy, a small rock-knoll — probably either a tor or small inselberg — rises through the peat. It carries crosses, an
incised slab, and a well dedicated to St Brendan, and I call it the Brendan 72
Rock-knoll. After the crosses had been erected peat continued to encroach
on the knoll and gradually built up to a thickness of 75cm around the 73
crosses, partially burying them. Recently they have been re-erected, and now stand clear of the peat. This partial burying is the clearest evidence of late growth of peat to be seen on the island. Smith in his eighteenth-century history of Kerry thought that the bogs of the area were still growing.

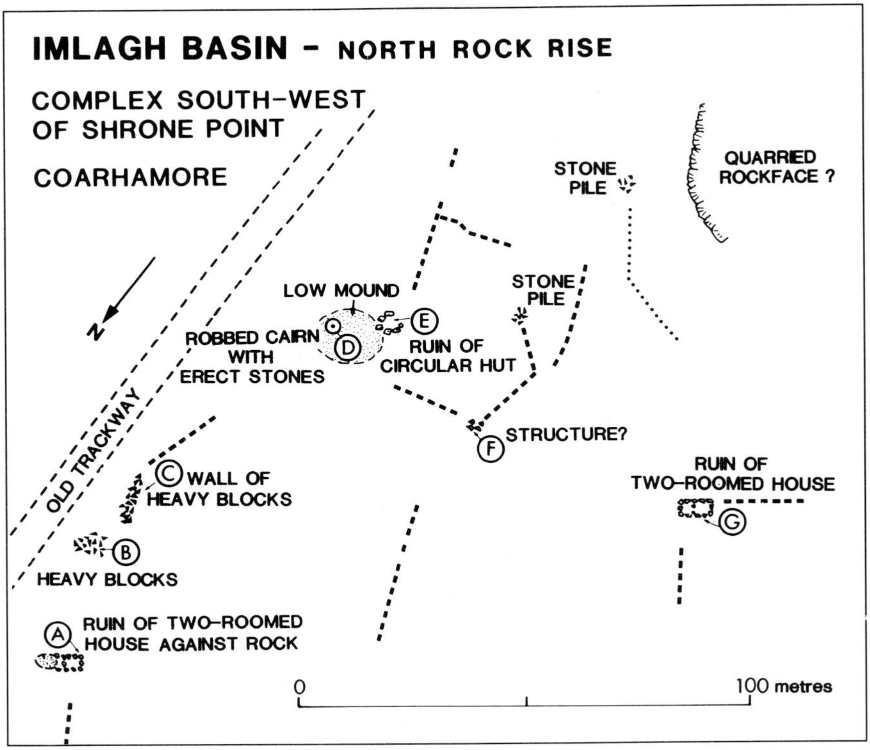

70 *Here there was a low rise on the northward rock slope, and over the rise the peat was thinner. Over an area with a radius of about 100m there were many man-made features, including lengths of wall. The old trackway from the road at Coarhabeg down to the fishing-site at Shrone Point crossed the east side of the Rise. A was the ruin of a two-roomed house against rock. B was a collection of heavy blocks. C was a wall of heavy blocks, like that at the Brendan Rock-knoll. D was a robbed cairn with erect slabs at its centre. E was the ruin of a circular hut. F was perhaps a structure at the meeting of some walls. G was the ruin of a two-roomed house.*

70 *Imlagh Basin, North Rock Rise, house against rock overhang.*
Here we are on the scarp-slope on the north side of the island, and rock overhangs are common. The east end of a two-roomed rectangular house has been tucked in under an overhang. The outlines of the house, which measured 5m × 4m, can be seen, and also the line of a cross-wall which divided it into two rooms of equal size.

Coastal Downland, Knockaunadagrean. 71
The cliffs in the foreground show the pale-coloured glacial gravels of the Downland resting on dark-coloured rock. The hillock of gravel, Knockaunadagrean, has been almost entirely quarried away; its remains are encircled by tractor tracks. Beyond the sward of the Downland we see the peat cuttings in the Imlagh Basin and in the Clynacartan Bog Embayment (extending towards the right-hand top of the picture). The tongue-like west end of the Main Escarpment protrudes on the left.

Imlagh Basin, Iron Age track (No. 18). 71
At this point a narrow track of earth and stone, about 60cm across and 35cm thick, was buried in the bog, with peat above and below it. A peat cutting with an arcuate face revealed a splendid diagonal section through the structure, which was running in a straight line. Peat at the level of the track had a ^{14}C date of 1700 BP, which places the track in the Iron Age.

72 *Brendan Rock-knoll, peat-buried cross.*
This picture, showing how a knoll of rock has become progressively buried by bog, graphically illustrates the continuing growth of peat on Valencia Island. A well and crosses were set up here in honour of St Brendan, possibly about AD 800. Since then 75cm of peat have grown up around the cross in the foreground. The highest point on the knoll lies behind the cross, and another slab stands there. All the rock is now covered by heath vegetation.

72 *Kildreenagh Dip-slope, ecclesiastical site.*
Used until recently as a cemetery, particularly for unbaptised children, this site has a long history. The presence of a cross incised on an ogham stone marks it out as an early ecclesiastical site. Towards the bottom of the enclosure we see two rectangular foundations, the larger on the left being perhaps those of an oratory. Across the enclosure are three circular foundations, probably those of 'beehive' huts where the monks would have resided.

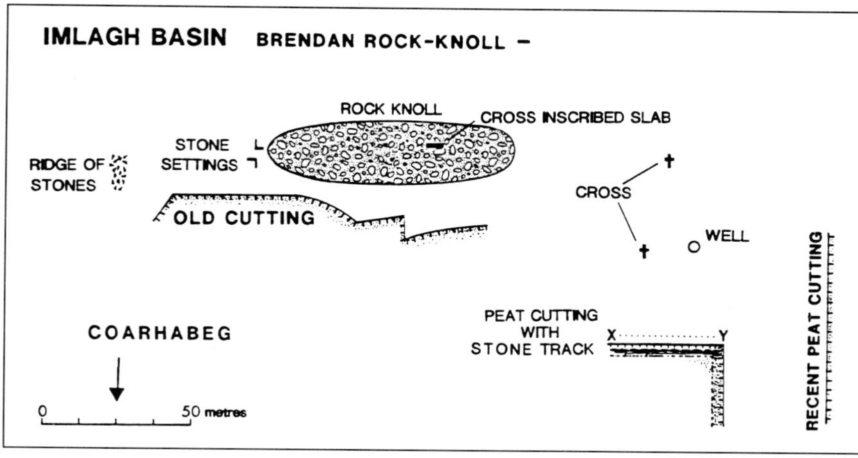

Only a small oval area of rock still rises above the peat; it carries a cross-inscribed slab. Immediately east of the knoll are two small stone settings. A little further on there is an obscure ridge of heavy stones; they are not the remains of a megalith, as Françoise Henry suggested. A holy well with an incised slab and two crosses are west of the knoll; until recently the crosses were largely buried by later peat growth, but they have now been re-erected. The peat cutting showing the stone trackway (No. 20) is a little further north. 73

Just north of the well, the end wall of a peat-cutting, 9m across, shows a continuous line of small cobbles. This is a diagonal cut through a stone trackway (No. 20), which appears in section in another cutting some 9m to the west, buried by 125cm of fibrous peat, where it is about 2m wide and consists of a single layer of stones. Where it is seen in diagonal section it is buried by 90cm of fibrous peat. A ^{14}C sample at the level of the track gave an age of 1720 years BP, an Iron Age date. Below the track was first 120cm of *Phragmites*-peat, then 20cm rich in wood, and then 15cm of muddy peat resting on mineral soil. The pollen diagram 15
shows that the locality was first wooded and then covered by grassland before it was buried by blanket-bog.

It is not easy to say when Christianity reached south-west Ireland. Tom Fanning considers that some sites go back to the fifth century. The well and the rough stone crosses are dedicated to St Brendan, a sixth-century saint. Such crosses are quite impossible to date satisfactorily; a ninth-century date is quite plausible.

25d Emlagh Bog Embayment 15ha Cool West

The area around the springs at the point where the beak-like projection from the Cool West Lobe forced the drainage stream to turn south-west subsequently became choked by peat, and I call the area the Emlagh Bog Embayment. This name *Emlagh* had already been used in earlier descriptions of archaeological finds here.

In 1984 a peat-cutting a short distance north of the springs showed an archaeological sequence. The uppermost layer was of fibrous 76
ombrogenous peat, and of this a thickness of about 50cm remained. Below 77

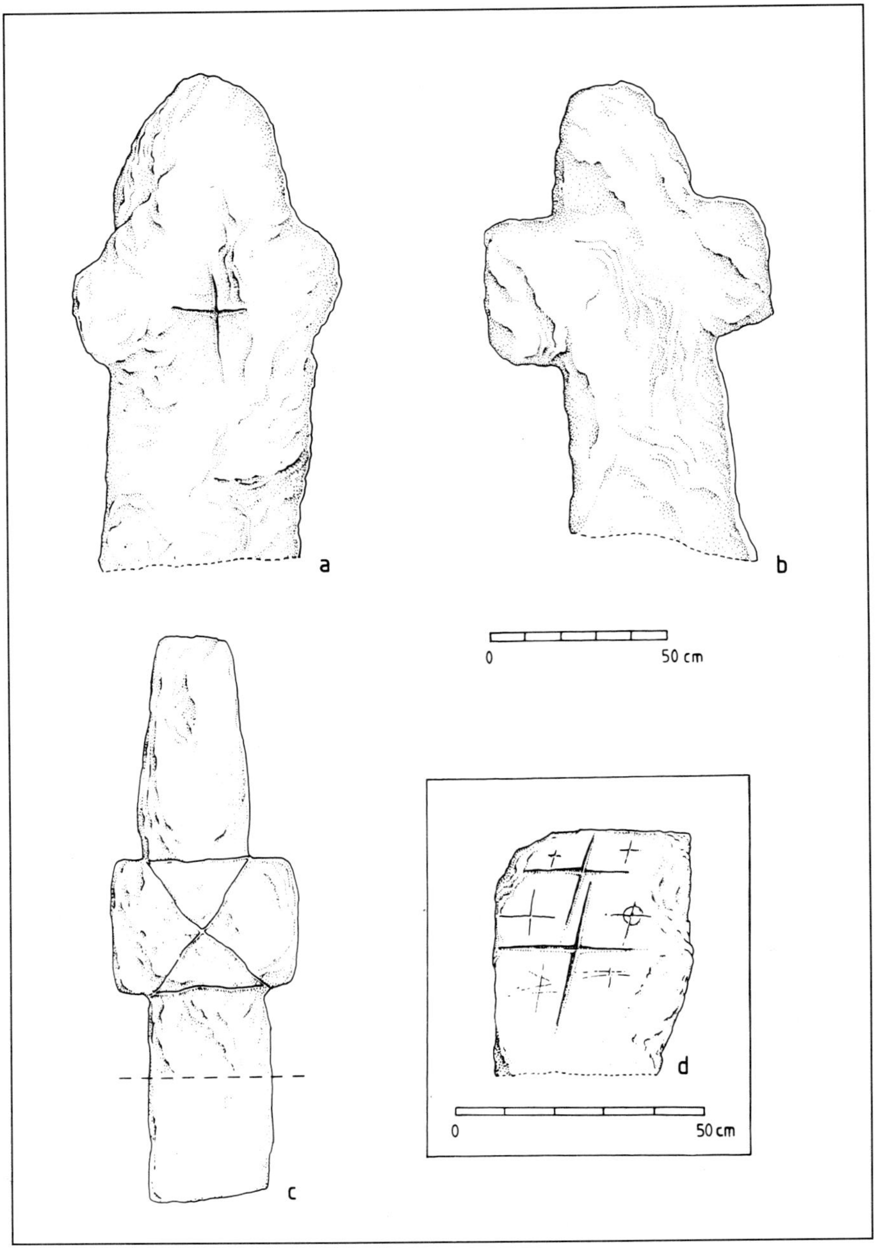

74 *a. Cross-incised slab on top of Brendan Rock-knoll.*
b. Cross, formerly largely peat-buried, near well on Brendan Rock-knoll.
c. Cross with incised saltire in Emlagh Bog Embayment.
d. Slab with incised crosses near well on Brendan Rock-knoll.

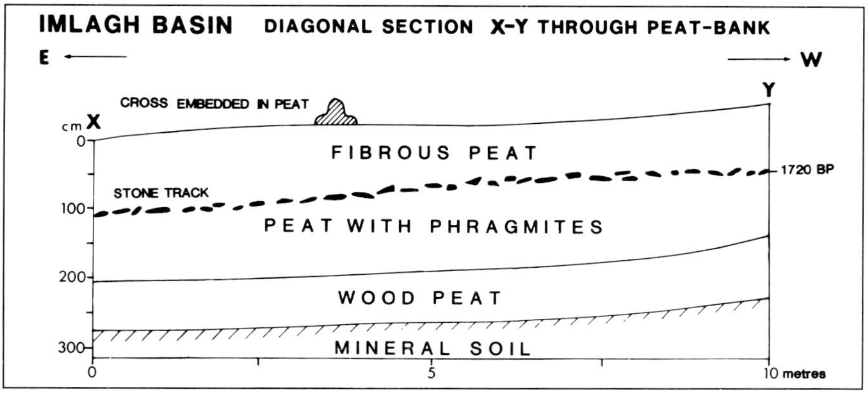

Brendan Rock-knoll, stone trackway. 75
The line of stones seen in the face of the peat bank are a diagonal section through a narrow track of stones laid on the bog surface about 1720 BP. More peat then formed.

this was a *Phragmites*-peat with a varying content of wood; it is about 90cm thick, and its base was dated to 5120 BP. Embedded in the peat from above downwards were (i) a layer of stones, dated by adjacent peat to 3930 BP (this feature is Bronze Age, and will be discussed below); (ii) a small oak stump in position of growth, dated to 4600 BP; (iii) a wall, essentially composed of two lines of large slabs leaning against one another, with in some places a gritty fill between them. There was about 20cm of *Phragmites*-peat free from grit and sand below the base of the wall. The base of the peat was dated to 5120 BP.

The bank was again cut for fuel in 1985 and 1987, and the diggers kindly left the bulk of the stones in place. By 1987 the wall was protruding diagonally from the bank for a distance of 2m, and its structure could be confirmed. There was a sheet of willow twigs at the level of the base of the wall, which suggested that the fen had been covered by scrub willow 89
when the wall was erected. Sheets of willow twigs were noted in other parts of the bog in the Basin, and pollen counts often produced very high values for willow. Willow twigs collected at the level of the base of the wall were dated to 4760 BP. This date is in accord with the date of 4950 BP for the agricultural phase at Reenarea Rise, 4790 BP for the grazing phase on Bray Top, and also with 4915 BP at Cashelkeelty.
Pollen counts from two sets of samples taken by Pete Coxon in the 18
immediate vicinity of the wall show agricultural disturbances at levels corresponding to the wall.

The layer of stones in the peat-cutting with a date of 3930 BP was 76
the oldest of the antiquities on the island that were placed in the Bronze
Age. When first seen in 1984 only a small cluster of stones was revealed; 77
cutting in 1985 exposed a horizontal layer of stones about 80cm across;
in 1987 the cutters again left the layer of stones in place as they cut back 90
the bank for a distance of 70cm. It thus seems reasonable to suppose that a trackway about 80cm wide and of a single layer of cobbles only

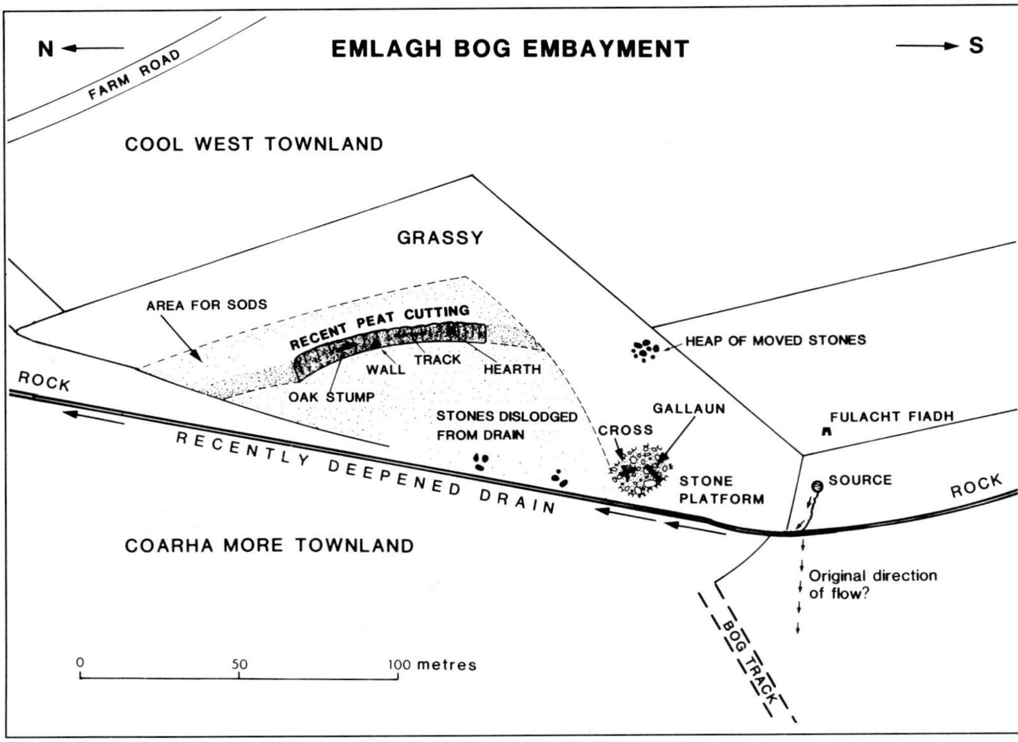

76 *There were springs in this vicinity, and before too much peat had built up the embayment was favoured by prehistoric man. This sketch-map shows the area near the springs (bottom right). A recent peat cutting showed Neolithic and Early Bronze Age features. In the course of cutting many stone slabs had been dislodged, and there must have been extensive structures buried in the bog. A stone cross and a gallaun were probably not in primary position. There was a small Early Bronze Age* fulacht fiadh *(3400 BP) close to the springs.*

was laid on the bog surface. It is remarkable that the stones of the trackway showed little signs of disturbance; this suggests either that the trackway was not much in use or that the bog surface must have been
18 relatively dry when the stones were emplaced. Pollen counts from the level of the trackway show typical agricultural disturbance.

Peat-cutting in 1988 revealed another feature 6m south of the track and at the same stratigraphical level. A number of flattish stones (about 30cm across) had been arranged to form a basin-like structure (about 50cm across). A thin (3cm) sheet of charcoal and stone fragments lay across the interior of the basin, and extended a short distance to the south. Probing suggested that the structure did not extend far back into the peat. It was perhaps a circular hearth of the same age as the trackway. There was no suggestion of pottery nearby, such as Seán Ó Ríordáin and I found at a somewhat similar hearth in bog at Rockbarton, Co. Limerick, many years ago.

A short distance east of the springs already referred to, an uncut bank

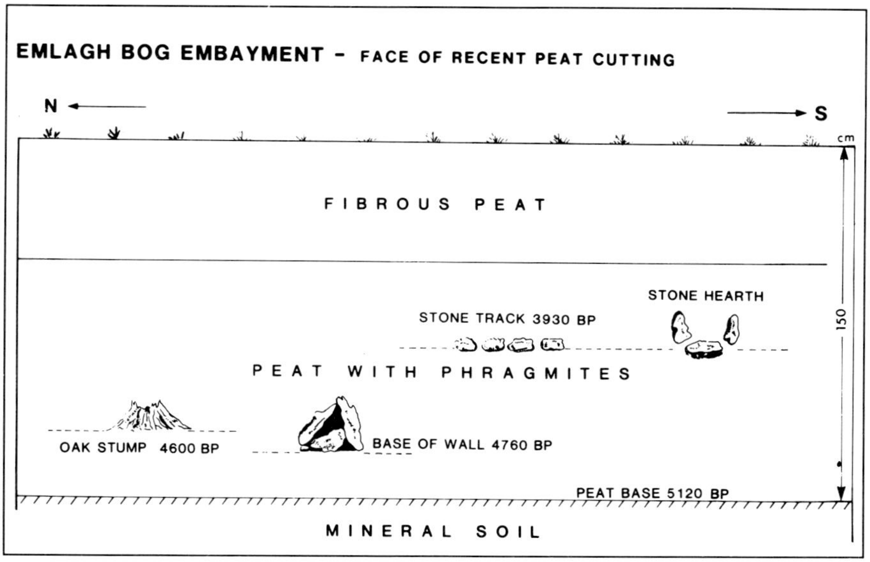

Sketch-section of face of recent peat cutting. 77
Peat started to form here about 5000 years ago when the area became a fen. Some 250 years later Neolithic people built a wall out into the fen. Later the bog surface got drier, and about 150 years later trees, including oaks, grew on the bog surface. Early Bronze Age man came to the site about 4000 years ago and laid a narrow track of small stones on the bog surface; he also had a fireplace nearby.

of peat, still 2m high, formed the boundary of a field from which peat had been removed. In 1984 a trench had been dug along the base of the bank, and this exposed two small *fulachta fiadh.* After a thin layer of peat had formed on the mineral soil, two small pits or trenches, about 5m across, had been dug. Low piles of upcast, about 3m across, rested on the basal peat. In the northern trench there was a thin layer of small stones and charcoal; the charcoal, which had a ^{14}C date of 3400 BP, was composed exclusively of twigs of ling, again suggesting absence of firewood in the locality. An older *fulacht fiadh,* dating to 3800 BP, on the Reenarea Rise had abundant charcoal of oak and other trees, suggesting that during their early years on the island Early Bronze Age folk cleared a lot of woodland.

In the vicinity of the peat-cutting there are several piles of large stones, presumably unearthed from the bog from time to time. Set in one of these heaps are a stone cross and two pillar stones; none of them are 74
now erect, but Westropp has an illustration which shows all three standing in a compact group. Legend connects the three with St Brendan. Françoise Henry was told that the cross had been found in the vicinity during peat-cutting, and was later erected here. The name 'Crush' (cross?) appears nearby on the six-inch OS map, but local people do not know the name. It seems most unlikely that any of the three were in their original positions when Westropp saw them. Westropp says 'the

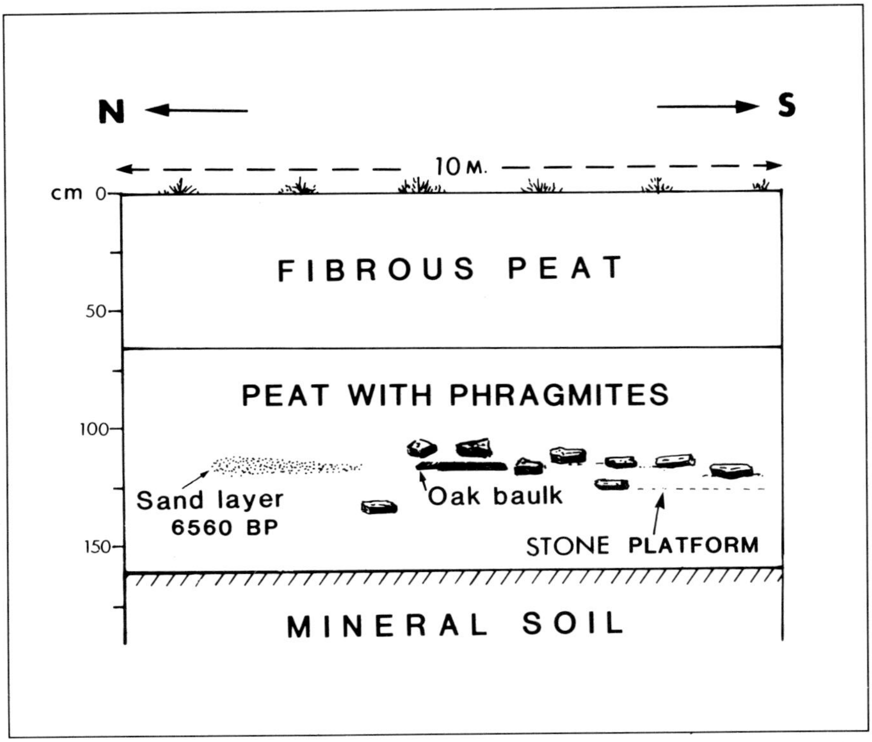

78 *Imlagh Basin, Clynacartan Bog Embayment.*
Sketch-section of peat face showing stone platform of Mesolithic age (6560 BP). Large slabs were laid on what was then the bog surface; a baulk of fossil oak (8910 BP) was incorporated into the structure.

cross is 4 feet 8 inches high, 13 to 9 inches across, and 20 inches over the arms; it has lines and saltires'. One stone was 6 feet 10 inches high, and Westropp considered it to be a gallaun; the second was 'a little pillar 3 feet high'. The cross in its material (slate), proportions and workmanship is remarkably similar to a cross in the graveyard at Kilmalkedar in the Dingle Peninsula.

35 **25e Clynacartan Bog Embayment 50ha Bray, Coarha Beg**
As the Imlagh Basin began to fill with peat, peat also began to fill the course of the tributary stream from the Clynacartan area. As it developed the peat buried a Mesolithic platform; in the nineteenth century the surface of the peat was cultivated.

65 At one point between the large rock-knoll already referred to and the
stream, at a low-lying point near the level of the stream, the wall of a
78 peat-cutting showed a sheet of flat slabs (each about 50cm broad)
extending for a distance of about 6m. A baulk of oak about 1m long
lay among the slabs. The slabs were laid on the bog surface after about

40cm of *Phragmites*-peat had formed, and a pollen diagram from the peat 12
suggested that it had started to build up about 7000 years ago. From the slabs a thin sheet of sand extended along the face of the cutting for some metres at the same level. A slice of peat containing the sand sheet was dated to 6560 BP, and this, a Mesolithic date, was taken to be the date at which the slabs were laid on the bog surface. The pollen diagram showed that the local woodlands had been disturbed by human activities.

The baulk of oak was also dated, and this gave a date of 8910 BP, considerably older than the building of the platform. Thus the timber was not cut from a living tree, but must be a piece of fossil 'bog-oak' collected by the platform-builders and used to strengthen the structure; it does give the date by which oak must have been established on the island. Peat growth continued, and a further 50cm of *Phragmites*-peat accumulated before blanket-bog started to form. Of this 60cm still remained, though the top of the bog had been cut away.

Large areas of bog surface in this Embayment were still intact in the
early nineteenth century. Today this surface shows ridges and furrows, 65
and some small 'gardens', relics of potato-growing in pre-Famine days. Some have thin layers of sand and fine gravel that were spread on the surface to improve soil quality.

25f Bray West Bog Embayment 65ha Bray

From the West Rock Rim a very gentle slope at first drained by a stream 52
led eastwards down into the Basin. Peat growth later crept westwards towards the Rim for a considerable distance, burying antiquities as it did so.

There are springs at the base of the Bray North Escarpment, and in 1984 a drainage trench running north to Coosagauna was cut through the peat of the Embayment. At three points in the trench small lenses of stone and charcoal, the remains of hearths, were intersected. Above one lens there remained about 60cm of humified peat without *Phragmites*, probably ombrogenous in origin. The lens was about 3m across and 50cm thick; a sample of ling charcoal from its base had a ^{14}C age of 3160 years, an Earlier Bronze Age date. Pollen of plants of cultivation
were associated with the hearth. Immediately below the lens was 80cm 17
of *Phragmites*-peat, which rested on mineral soil. The basal layers of the peat also showed agricultural activity, and there was some content of tree pollen; this disturbance phase is probably Neolithic in age.

There were a number of old walls in this area. A short (75m) narrow
bank (No. 1) cuts off the tip of Beenanniller Head; a second slight trench 52
is perhaps modern. A long (475m) ill-defined bank (No. 2), best seen in aerial photographs, runs south from the foot of the Bray North Slope to the west side of Coosagauna. A modern bank runs along the foot of the Slope. A short length of curved wall (No. 3) survives above the south side of the bank; the rest was cleared away when the bank was built, and fields were laid out on its north side. Three stone walls (Nos 4-6) run down the Slope. Wall 4 is a short (100m) stretch of stone wall running

diagonally down the Slope to the modern bank, and after a gap it reappears on the same alignment for 50m on the east side of Coosagauna. Wall 5 is a short stretch of stone wall also running diagonally down the Slope; it cuts through the curved wall (No. 3) to be in turn cut off by the modern bank, and reappears in the same alignment below peat for 5m north of the road. Number 6 is a short length of stone wall running down the Slope 200m east of No. 5; it is not seen beyond the modern bank. Wall 7 is a short (100m) double stone wall running east from the east side of Coosagauna out into bog.

Further to the east there were two problematical features. Number 20 was a length (300m) of bank running south-east from the tributary
52 stream to the bend in the road leading to Clynacartan, with which it was in alignment; it has been demolished by a land improvement scheme. Number 21 is enigmatic; it is a suggestion of two short (75m) parallel banks north-east of the tributary stream.

25g Cool Bog Embayment 15ha Cool East and West

61 As already noted, a small stream flowed north from the hollow between the Cool East and the Cool West Lobes, down to the sea at Coosheenadagallaun. Peat formed here also, but no antiquities were seen in the Embayment.

THE SOUTH-EAST SLOPE

26 Dysert Slope 190ha Coarha Beg, More

A remarkably even, gentle slope (about 5°), facing south-south-east, runs from Coarha Top and its western ridge down to the sea. In the centre of the Slope there is a small stream in a very gentle valley. To the east of the stream, and for a short distance to the west, the solifluction slope runs down to sea-level. The coast then turns north-north-west, and rocky cliffs begin to rise. Most of the lower slope was formerly peat-covered.

There is a gallaun on the upper edge of the Slope where it meets the Coarha Top, and a ploughed-out rath near its eastern limit. Just west of the stream and not far from the gallaun there is a Children's Burial Ground,whose margins have been trimmed by farming operations. Apart from numerous grave-marker stones it is featureless, except for a short line of erect slabs in the south-east corner.

In the valley, where some peat survives, there is a low round earth bank about 12m in diameter, and also a round bank about 11m in diameter, which appears to have been cut out of peat.

From the cliffed shore at the west end of the Slope a narrow tongue of rock projects westwards. Here there is a seventeenth-century star-shaped fort, which may have been built on the site of an earlier promontory fort.

4 **27 Oonagranshee Lobe 90ha Coarha Beg, More**

This very shallow smooth lobe, inclined to the south-east, starts from the east end of the Coarha Top and runs down to the sea. Soil drainage is good in the upper part of the Lobe but deteriorates rapidly downslope, where some of the poorest soils on the island occur. The lower slope was formerly covered by peat. On the east side of the Lobe, at an altitude of 30m, after the peat had been cut away no attempt was made to reclaim the land. Lower still on the same side the ground is covered by a very wet *Molinia*-meadow; this is growing on an impermeable homogenised mixture of glacial debris; no reclamation has been attempted.

On the upper part of the Lobe there are the damaged remains of an oval enclosure, originally perhaps 45m across. Inside is a ruined rectangular drystone house, 10m x 7m; a ruined drystone circular hut, 6m in diameter, with a collapsed souterrain below; a souterrain chamber; *81*
two thin erect slabs; and traces of other ruins. It is recorded on the six-

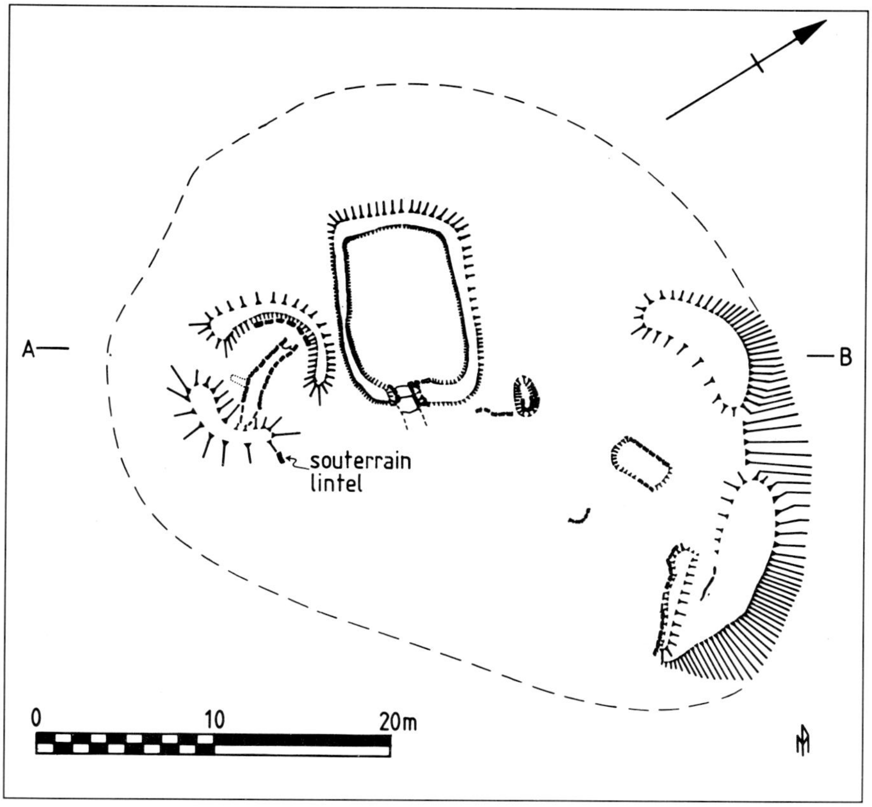

Oonagranshee Lobe, complex house site. *81*
This complex site has been very much damaged, but it may represent a substantial farm of the Early Christian period. There was probably an encircling bank, with a substantial rectangular house and minor buildings inside it. There were two souterrains; one was below the floor of a circular hut—an arrangement which also occurred at the Cool West Lobe; the second, which was intact, was a short wedge-shaped structure. Traces of other ruins suggested further structures.

inch OS map as 'Oonagranshee Cave'. The site appears to have affinities
42 with the two house groups on the Coarhabeg Lobe. Oonagranshee lies
43 to the east of Coarha Top; the other sites lie to the west of the Top in analogous situations. The sites are 2km apart. There is also a ploughed-out rath on the Dysert Slope, 500m to the west.

28 Reenarea Rise 15ha Coarha More

This is an oval rise, no more than 15m high, surrounded by the sea on three sides, and bounded on the north by a gully between the Rise and the base of the slope on the western side of the Oonagranshee Lobe.
83 It is about 600m long, 200m across, and aligned north-east/south-west. This is a remarkable unit in that it breaks the uniform slope of soliflucted material that marks most of the south side of the island. Though of low elevation, it is a rock rise, with an identity of its own. At first the soil was well drained and of good quality, but the Rise subsequently became completely buried by peat, and the attractiveness that it had had for settlement steadily diminished.

In an area of cutaway bog 85cm of deposit still remained, exposed in the side of a recent drain. From above, there was 30cm of highly humified peat, then 15cm of charcoal and burnt stones — the margin
83 of a *fulacht fiadh* — and then 40cm of woody *Phragmites*-peat on mineral soil. The base of the peat had a ^{14}C age of 5630 years. The charcoal was not of ling as at the other cooking-place sites, but of various woods — oak, poplar/willow, and hawthorn. To avoid the possible inclusion of charcoal from old 'bog-oak', the ^{14}C sample was of small-vesselled wood only, and this gave an age determination of 3800 BP. There was a small piece of uncharred yew wood in the peat immediately below the burnt materials, and this had a ^{14}C date of 3850 BP. The *fulacht fiadh* is thus Earlier Bronze Age, and of the same age as the stone track in the Emlagh Bog Embayment. The fuel for this site had been collected in the surrounding woodlands; the fuel for the slightly later sites noted earlier had only charcoal of ling, suggesting extensive woodland clearance by Earlier Bronze Age people between 3800 and 3400 BP, a period of four hundred years.

Examination by Pete Coxon of the pollen in the peat below the hearth debris produced a still greater surprise. Counts from the base of the peat
14 showed the presence of high forest of birch, hazel and oak, with ivy and some pine. There was then a short phase of woodland clearance and cultivation, dated by ^{14}C to 4950 BP, a phase which must be Neolithic in age, and compares well with the dates for the peat-buried wall in the
89 Emlagh Bog Embayment (4760 BP) and the mor on the top of Bray Head (4790 BP). Cultivation was abandoned, and willow and birch quickly invaded the deserted fields. There was still woodland in the area when the Earlier Bronze Age people arrived.

If the peat below the *fulacht fiadh* accumulated at a constant rate between 5600 and 3850 BP, then the pollen samples are separated from one another by a time interval of 160 years. The clearance phase is recorded in two samples only, and therefore it probably existed for some 300 years.

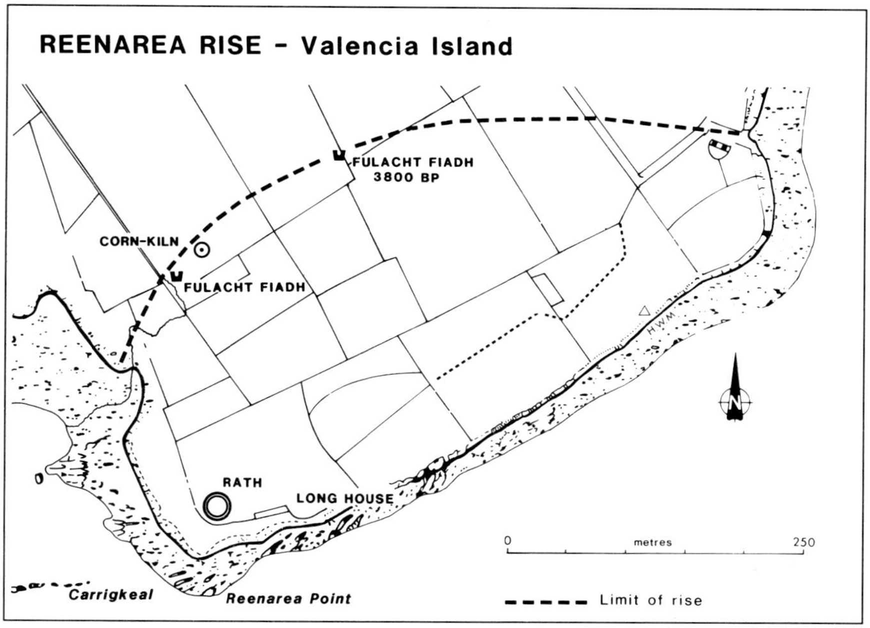

This small, oval rock rise has many remarkable features. On the north, in a peat-filled gully which separates it from the Oonagranshee Lobe, there is a mass of stone and wood-charcoal derived from an Earlier Bronze Age fulacht fiadh *with a* ^{14}C *date of 3800 BP. On the north-west, tucked in against an irregular rock-knoll, are the remains of what was probably an Early Christian farm, dating to the fourth century AD, with a* fulacht fiadh *and other remains. There are also a rath at the south-west corner and nearby the ruins of a long house, perhaps relatively recent.* *83*

A corresponding phase near Cleggan in Connemara is thought to have lasted about 200 years.

At the western end of the Rise, where bog was giving way to rocky ground at the bottom of the Oonagranshee Lobe, there was another *fulacht fiadh* on peat which was excavated by the South-West Kerry
Archaeological Survey under John Sheehan. Four large erect slabs at *84*
right angles forming the walls of the trough, and another slab providing
a floor, had been partly sunk in peat, and other slabs had been thrown *85*
down around. A stone spindle-whorl of Early Christian style, decorated
with an incised circle on each face, was found near one of the slabs. *90*
A large mound of burnt stone and peat charcoal lay nearby. It was decided not to attempt to date the peat charcoal; a ^{14}C sample from the uppermost layer of the peat below the mound had a date of 2950 BP.
The mound yielded pieces of a hand-moulded grass-marked pottery *85*
vessel, probably dating to the fourth century AD.

About 30m east of the cooking-place there was a feature which
suggested either a circular hut with a souterrain or a drying-kiln with *84*
an extended flue. There was a short length of drystone walling nearby. Excavation here would probably be rewarding.

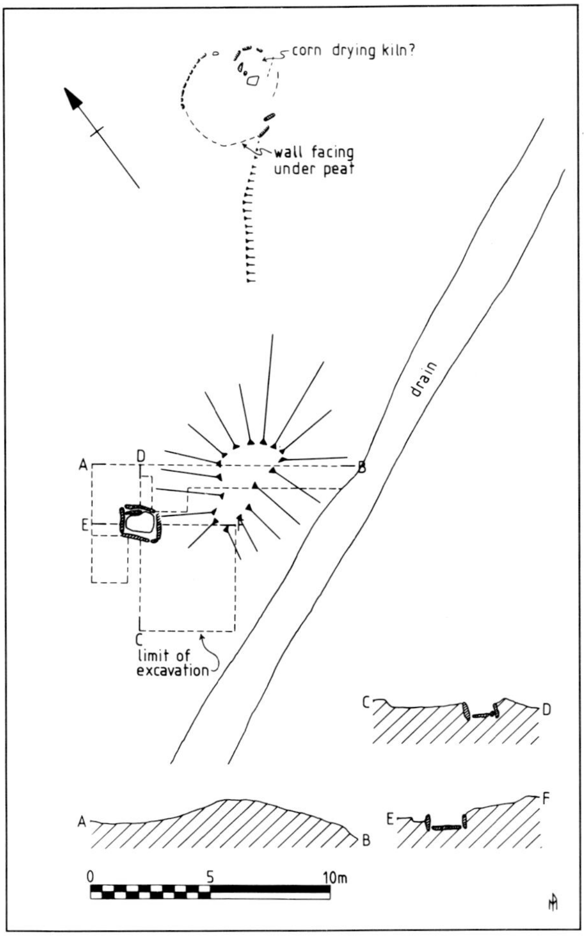

84 *Reenarea Rise, Early Christian site. At top, plan of stone features, including kiln(?). Below, plan and section of* fulacht fiadh.

At the south-western point of the Rise (Reenarea Point) there is a rath, 30m in diameter, with the sea to west and south. In such a position the rath takes on something of the character of a promontory fort. Its waterlogged ditch probably holds organic occupation debris.

Towards the west end of the Rise there are cultivation ridges on the peat surface. There are also traces of walls under peat.

At the edge of the low cliff immediately east of the rath are the bases of stone walls which once formed a long row of rectangular houses; these may not be of great antiquity. The site would have provided a good base for smuggling operations.

29 Donnybrook Valley 135ha Coarha More, Tinnies Lower
A well-defined, shallow, parallel-sided valley, reaching back almost to the Spine-ridge, and falling east-south-east to the coast; it was formerly peat-filled to a height of about 60m, and a considerable amount of peat still remains. On its eastern side the Valley has some improved ground but today it is almost derelict, giving only some rough grazing.

No prehistoric features were seen, but near the floor of the Valley some old cultivation ridges and 'gardens' lie on top of the surviving peat.

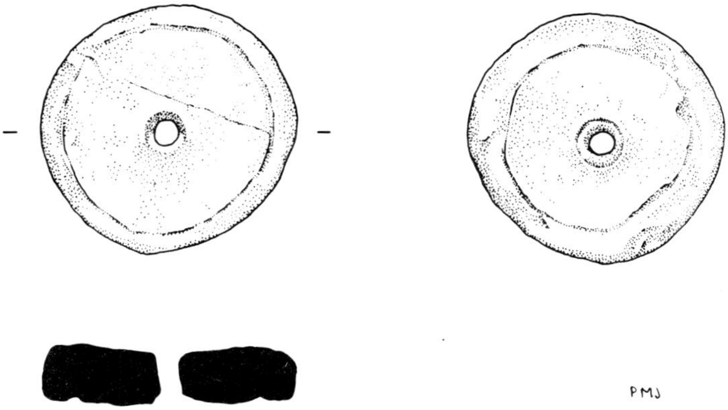

Reenarea Rise, fulacht fiadh.
Decorated spindle-whorl of Early Christian type (1/1). *85*

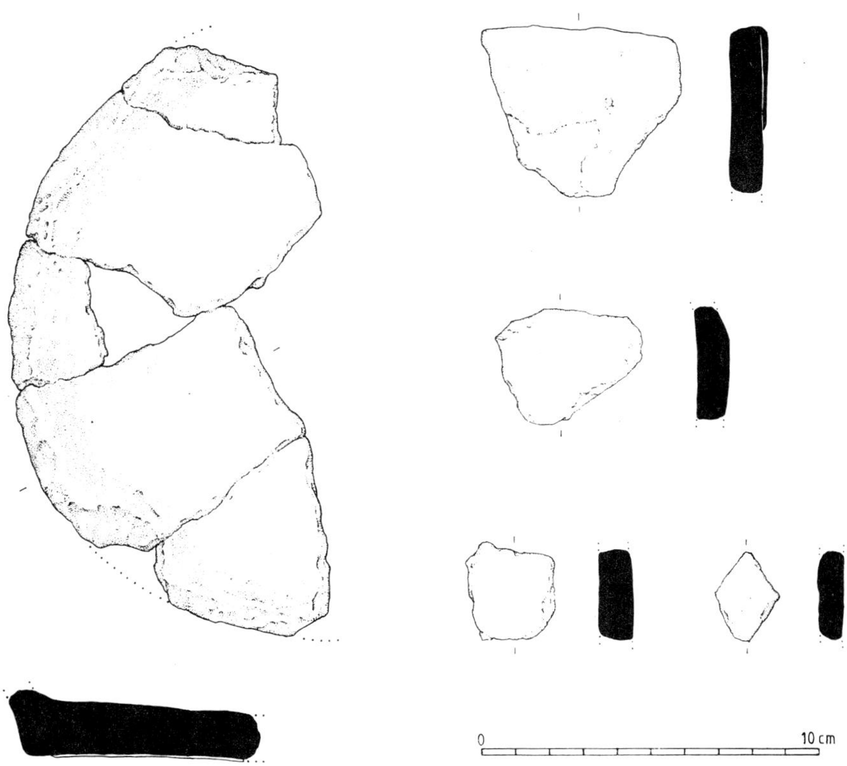

Reenarea Rise, fulacht fiadh.
Sherds of hand-moulded grass-marked pottery, possibly fourth-century AD. *85*

4 **30 Tinnies Lobe 150ha Coarha More, Tinnies Lower and Upper**
This Lobe springs from a small flat stretching out from the side of the Kildreenagh Slope, and drops down to the south-south-east. At first the fall is interrupted by short low rock ridges across the Lobe, but then becomes smoother and more gentle, with low downward slopes to either side. It was probably formerly covered by peat to a height of 60m. The upper slope with good soils is now largely covered by improved grassland.

High on the Lobe at 100m there is a large gallaun, 3m high. One clachan is sited at the top of the Lobe. A low knoll at about 70m further
114 down the crest of the Lobe was the site of a small clachan, which may well have been built at the upper limit of peat growth.

31 Chapeltown Valley 150ha Tinnies Lower, Ballyhearny
A funnel-shaped valley — almost a basin — starting on the north below the Kildreenagh Slope and the Gortgower Saddle, with the Tinnies Lobe to the west and the Ballyhearny Lobe to the east, it is drained by two streams, one from the north, one from the north-west, which meet at an angle of 45° and a height of 25m at the site of the present village. Their joint channel narrows to a throat between the lower ends of the two lobes, and then widens slightly as it approaches the sea. It is difficult to understand the erosional processes by which the valley was shaped; similar land forms can be seen on the south slopes of the Slieve Mish Mountains in the Dingle Peninsula, also of Old Red Sandstone. In the low-lying ground immediately above the confluence of the streams some peat still survives; wood of pine and yew can be seen at the base of the peat. Subsequently peat extended upslope to a height of about 70m. Today most of the basin is occupied by poor wet land, mostly in small rectangular fields.

Except for one gallaun at 100m on the eastern margin of the valley, no antiquities were noted. The church that gives the village its name was built in 1819.

4 **32 Ballyhearny Lobe 105ha Ballyhearny West**
A subdued lobe with axis sloping south-south-west and reaching almost to the mouth of the Chapeltown river, a short distance inland from its mouth; it was probably formerly peat-covered up to a height of 70m.

No antiquities were noted.

33 Laharan Dip-Slope 240ha Ballyhearny East, Laharan, Kilbeg East

This is a concave slope, steeper above (about 8°) as it drops away from Kilbeg Top, and flatter below as it runs down to tide-level. Below the upper road from Knightstown to Chapeltown, which is remarkably level at about 70m, the slope is about 4° on a slaty soliflucted mixture of rock and drift, which has very poor drainage. This part of the Slope was formerly covered by peat. Writing in 1813, Nimmo reported that 424 Irish acres (about 300ha) of bog remained on the harbour side of the island. The lower road along the shore did not exist in 1842 when the first OS survey took place, and at that time a substantial amount of peat may have remained at low levels. Several narrow tracks run down the slope to the shore and these were probably used for drawing seaweed, as Estyn Evans has noted for County Down. All the lower slope is now improved grassland, and many old field boundaries have been swept away.

At high-water mark on the shore 500m west of Reennaloughan Point there are the remains of a *fulacht fiadh,* which was later damaged by running water. Some 40cm of peat, with *Phragmites* at the base, rest on glacial material. A layer of grey sand 10cm below the top of the surviving peat expands laterally into fragmented slate and quartz, with a content of burnt stones and charcoal; the charcoal had a ^{14}C date of 2970 BP, which places the cooking-place towards the end of the Earlier Bronze Age. A few metres to the east burnt stone debris was more conspicuous.

High up on the Slope, just where it is about to give way to the Kilbeg Top above, and not far below the wedge tomb on the Top, there is a
remarkable structure, 25m in diameter, surrounded by a double wall *88*
of vertical slabs, the inner and outer faces being 1.5m apart. It may have been a cashel, or it may be a type of double stone circle, a monument style known elsewhere in south-west Ireland.

Above the break in slope there are two holy wells. *Tober naom trigion* is a simple slate-lined well in a 'garden' with cultivation ridges.
Tobereendoney (St Dareca's Well) is a pretty, drystone, corbelled structure. *88*
Both are still venerated.

There is a Children's Burial Ground at the break in slope with many paired erect small stones and one stone pile. There is a fragment of a curved drystone wall, perhaps once part of quite a large building. The townland name is Ballymanagh — the place of the monks; another nearby townland is Kilbeg. Is this the site of the missing church of Kilbeg?

East of Reennaloughan Point are the ruins of Revenue House and its associated cottages. Smuggling and piracy were rampant along the south-west coasts in the seventeenth and eighteenth centuries. As the Channel had two exits a revenue cutter, stationed near the Point, was well placed to give chase.

The ruins of Zelva School, built by the Erasmus Smith trustees at the end of the eighteenth century, are a prominent landmark on the higher ground at the east end of the Slope.

34 Knightstown Nose 130ha Farranreagh

A lobe of the 200ft contour projects north-east below the lower slopes
of Kilbeg Top. If we take a line from there to the north-eastern corner *35*
of the island where Knightstown now stands, we have a slope of about 5°. From that ridge the ground slopes gently away on either side down to sea-level, giving a nose-like projection into Valencia Harbour. Above the shoreline just south of Knightstown the slope is 2°. Peat later formed on the Nose up to a height of about 65m. This has been cut away and

88 *Laharan Slope, circular bank retained by upright slabs.*
At first sight this circular monument looks like a rath or cashel. But its bank, both on its inner and its outer side, is retained by a ring of erect slabs of stone, and its purpose might have been ritual rather than residential.

88 *Laharan Slope, holy well.*
This drystone corbelled structure has been built over a spring facing south on the Laharan Slope. It is called Tobereendoney *and is dedicated to St Dareca, the patron saint of the island; it is still venerated.*

Emlagh Bog Embayment, Neolithic wall.
The wall was erected by pushing a double line of slabs down into already-formed peat, and filling the space between the slabs with earth. Mineral soil below the bog can be seen to the right of the base of the larger stone, and the earth between the stones sits on what was the surface of the bog at the time the wall was erected. The surface had a ^{14}C date of 4760 BP, a typical date for Neolithic man in Ireland.

90 *Emlagh Bog Embayment, Bronze Age track.*
A few metres south of the Neolithic wall a line of small stones projected from the peat face at a higher level. Later peat-cutting showed that the line was a cross-section of a narrow trackway. Peat at the level of the track had a ^{14}C date of 3930 BP, so the track must belong to the beginning of the Bronze Age in Ireland.

90 *Reenarea Rise, Early Christian* fulacht fiadh.
The fulacht fiadh, *which lies on peat, is at an early stage of excavation by John Sheehan. In the centre we see the slab-lined cooking-trough. On the right we see slabs lying on peat; a decorated spindle-whorl was found near one of them. On the left we see the exposed face of the mound of stone and ashes that built up progressively as the trough was cleared of debris; it yielded pieces of pottery, perhaps of the fourth century AD. The surrounding area is buried by peat.*

the area turned to agricultural and residential use. The 'reagh' (*riadh*) element in the townland name implies cold or grey soil, a very correct description.

In view of the very short distance from the point of the Nose to Reenard on the mainland (less than 1km), this is an obvious location for a ferry.
As early as 1830 there was a settlement here called 'The Nose', but the 35
area was completely transformed when the Knight of Kerry laid out the village of Knightstown in 1840.

Nothing of great antiquity can be traced, but there are two Children's Burial Grounds on the lower slopes south of the village, which may be later than the removal of peat.

35 Kilmore Shelf 30ha Farranreagh, Glanleam
North-west of the Knightstown Nose there is a small shelf, gently sloping
to the east-north-east, which is cut back into the east side of Kilbeg Top, 35
and ended on the north-west by the edge of the Glanleam Pocket. Its geomorphological development is not clear. Given its aspect, its upper end may have held a snow-pocket which cut back into the Top; the slope of the cut is 15°.

Except for the cemetery underneath the scarp, the shelf is covered by farmland. There certainly was a church here at the beginning of the seventeenth century, but there is nothing of great antiquity to be seen in the cemetery today. Downslope are the ruins of a nineteenth-century Church of Ireland church, which incorporates an older tower.

THE EBB AND FLOW OF THE ENVIRONMENT AND MAN

The environment certainly did change on Valencia Island: forest gave way to open country, ground that had formerly been dry became covered by bog. It is not easy to say exactly what type of landscape confronted the successive waves of settlers who tried to wrest a livelihood from its limited resources.

The total area of the island is about 3000 hectares. At the time when bog on the island was at its greatest extent, say about AD 1800, it probably buried about 500ha in the Imlagh Basin and about 1000ha on the south-east slope, thus covering about half of the island. In the other half about 600ha (20%) were of good agricultural land, and about 400ha (13%) were of fair agricultural land. The remaining 500ha (17%) were of rocky or poorly-drained land, only fit for the roughest of grazing.

And these proportions did not remain constant. On the north shore the Coastal Downland is built up of glacial gravel. In early post-glacial time the soil was probably well drained and covered with woodland. After the trees were removed, perhaps by Earlier Bronze Age man, they were replaced by grassland of fair quality on a soil whose drainage was still good. This grassland may have been exploited by Late Bronze Age people. But then, perhaps owing to soil exhaustion, a deterioration in climate, or burning, mor developed and blanket-bog vegetation began to creep upslope. As it formed, the peat not only buried the soil but also impeded the drainage of the soil further upslope, not yet covered by peat. As the bog expanded, the area available for grazing was reduced.

Today the downland soil is a gleyed peaty podzol covered by a thin layer of peat which carries not grassland but a plantain-sward of very low feed value. Animals do stray over it, but with very much less profit than that of the Bronze Age beasts, who not only had better fodder but a much greater area to roam over.

In the Neolithic, livestock grazed on pastureland on the top of Bray Head, and probably on Geokaun Top and Kilbeg Top as well. Today the Tops are covered by blanket-bog, which offers only the poorest of grazing.

Man's changing agricultural practices also had profound effects. Prehistoric peoples introduced cultivation, nineteenth-century folk covered almost the whole island with potatoes, and today grassland is king. All these changes altered the environment.

What would happen to the environment if man's influence were removed? Would the woodlands return? Trees will still make an effort to grow, if encouraged to do so.

On the south-eastern side of the island on the Laharan Slope, immediately beside Ballyhearny House, a rill runs south down the hillside in an exposed situation. Some time after 1842 the owners of the house planted a strip of woodland, 1km long and 100m wide, along the course of the rill from a height of 70m down to sea-level. Judging from their frequency, alder, ash and sycamore were the trees planted. Today the greatest trunk diameter is about 50cm, and about 15m is the greatest height reached. Owing to wind shear from the west, almost all the taller trees have broken tops; it would be interesting to know whether all were broken at the same time by a severe gale, or whether breakage is an inevitable consequence of attaining a certain height. There are a number of dead trees.

There is quite good ground cover, with blackberry, blackthorn, bluebell, celandine, golden saxifrage, hawthorn, Irish spurge, ferns, mosses and liverworts common. Honeysuckle and ivy also occur.

Other trees were rare, and I only noted beech (6 specimens), birch (4), elm (occasional), holly (2), oak (1), and pine (2). The only oak seen was a stunted young tree. There was no hazel, and I did not see any hazel anywhere on the island.

Much of the Glanleam Pocket has dense woods, with many exotic trees. Rarities are fewer on the north-facing fault-scarp; here ash and sycamore are dominant and alder is rare. Beech and pine also appear in the canopy. Hawthorn, holly, whitebeam and willow occur at lower levels, and there are also honeysuckle and ivy. I saw one elm and one oak sapling, about 3m high. I did not see any birch.

At the head of the Kilmore Shelf there is a hollow which gives some shelter. Beech and exotic conifers were planted here at some date, and the trees have made fair growth; because of grazing there is little undergrowth. There are some alder and sycamore, and I saw one ash, two birches, and three fair-sized oaks, probably planted.

These limited observations suggest that in the absence of human activity there might be a slow re-expansion of woodland. But what trees would compose it? Sycamore would clearly be to the forefront, but what about those that were formerly common on the island? Would hazel return? Would birch spread again? And what about oak? It seems extraordinary that the tree which gave the island its earliest name — *Dairbre* — should be so rare today and show only signs of feeble growth.

Prehistoric Valencia

MESOLITHIC

Our knowledge of this period is very slight, but nonetheless there must have been Mesolithic people on Valencia in sufficient numbers to have had some influence on the woodlands which then clothed the island.

At one site in the Clynacartan Bog Embayment we have a slab structure of this age. Here there is exposed in a cutting in fen-peat a
78 line of horizontal stone slabs which tails out laterally into a thin layer of sand; peat surrounding the sand layer had a ^{14}C date of 6560 BP. A baulk of oak among the slabs had a date of 8910 BP. As the contemporary fen-surface was probably very wet, it would seem that some type of boat or sledge must have been used to move the stones out into the fen.

This slab structure must be Mesolithic in age. The date is later than that for the first arrival of Mesolithic man in Ireland (about 9000 BP) but older than the date (5500 BP) for the kitchen-middens at Ballyferriter, 30km to the north on the Dingle Peninsula.

In Mesolithic times the lower parts of Valencia Island were densely covered by forest. The population was probably very small, but some
12 limited clearances were made in the woodlands, as pollen of plantains, sure indicators of open ground, was recorded in the counts; charcoal was also present. Where the ground was poorly drained, soligenous fen-peat was forming.

NEOLITHIC

For the Neolithic period on the island we have three lines of evidence, from pollen diagrams, from radiocarbon dates and from a structure.

Contained in the diagram from Reenarea Rise, which runs from 5630
14 BP to 3800 BP, we have a classical example of Neolithic land clearance, about 5000 BP. Oak and hazel are cleared away, and grasses and plants associated with cultivation take their place; tree pollen no longer dominates the air, and pollen of heather, which is growing nearby, has opportunity to be recorded in the counts. The clearance plots are then abandoned, birch and willow followed by oak and hazel invade them, and grasses and heathers no longer make a significant contribution to the local pollen rain. If we assume that the peat accumulated at a constant rate during the 1800 years covered by the diagram, then this clearance phase lasted about 350 years.

Though its opening date is rather earlier we see a very similar picture

at Cashelkeelty, where clearance patches are abandoned about 5000 years 13
ago and forest with oak, alder and birch builds up once more. Such a
clearance programme may have been standard practice with the first
Neolithic farmers to reach western Ireland, as a diagram by Karen
Molloy and Michael O'Connell from the vicinity of Clifden in
Connemara shows a clearance phase followed by woodland regeneration.
From their detailed work they suggest that the clearance lasted for about
two hundred years. Cereal pollen was very rare in Connemara; none
was noted in Valencia.

In the Emlagh Bog Embayment a number of features are exposed
in a peat bank where the base of the peat was dated to 5120 BP. A short
distance above the base was a stone wall built of a double line of slabs 77
with a fill of stony earth between; the wall was dated to 4760 BP, a
perfectly acceptable Neolithic date. Above again were an oak stump in
position of growth (dated to 4600 BP) and a stone trackway (dated to
3930 BP, a Bronze Age date). Pollen counts from two sets of samples
taken by Pete Coxon in the immediate vicinity of the wall show typical 16
agricultural disturbances at levels corresponding to the wall. On Bray
Top the base of the peat had a ^{14}C age of 4790 years, and pollen counts 18
showed that the area was then covered by weedy pasture. Apart from
these three sites with radiocarbon datings, the pollen diagrams from three
other sites, the Brendan Rock-knoll, the Bray West Bog Embayment 15
and the Imlagh Basin, also indicate Neolithic activity. 17

At the Imlagh site a slab trackway dated to 3930 BP lay on 160cm 18
of fen-peat; there was no mor at the base of the peat. The basal 10cm were sampled to try to determine the date of the initiation of peat formation. A range of tree pollen, including pine but not alder, was present at low values. When the peat started to form, the vegetation on the slope above was probably open. Dominant grass pollen with some weed pollen showed that there was weedy pasture in the immediate vicinity, but this was quickly superseded by quantities of ling pollen — *Sphagnum* spores also increased greatly — showing that a wet heath had replaced the grassland. Having regard to the depth of peat that had formed before 3930 BP, the grassland phase is probably Neolithic in age.

Neolithic finds dated to the millennium 5000 to 4000 BP are well known in the west of Ireland and at Lough Gur in Limerick, and it is becoming clear that these folk must also have pushed into the less fertile lands of south-west Ireland. If their farming was centred on grass, the long growing-season of the south-west may have attracted them. A Neolithic flaked stone axe was recently found on Beginish, close to Valencia, and a second on the mainland at Garrydine, 16km to the north-east.

In Neolithic times the lower slopes were still covered by forest, but the small population did open up areas where they farmed for a short period, creating fields. When a clearance was abandoned, trees quickly invaded it. On the uplands, in areas where heather growth was dense, judicious burnings would encourage the growth of grass and provide extensive grazing. On valley floors fen-peat continued to form.

EARLIER BRONZE AGE

99 Only one adequately documented object of Bronze Age date is known from the island, a Late Bronze Age sword, now in the National Museum. There is a nineteenth-century record of a gold ring, which might belong to any phase of the Bronze Age, and a spearhead, still with its wooden shaft, which might go back to the Urn phase of the Bronze Age. Neither is known today.

From the mainland nearby there are three records. The oldest is of a hoard of flat copper axes, presumably from the Beaker phase at the opening of the Bronze Age, from the vicinity of Cahersiveen. The National Museum has a wing-flanged bronze axe with a stop-ridge from the bog at Emlaghlea on the mainland, 10km to the south-east; this type of object is elsewhere in Ireland associated with Urns, and might be placed before 3000 BP. At Pound, across the Portmagee Channel at a height of about 100m on the hillslope looking north to Valencia, a cist was recently destroyed when a slope of glacial gravel was being 'improved'. Rose Cleary rescued one piece of a Pygmy Cup, a type of vessel usually
71 associated with Urns. On the island, on the Coastal Downland, an isolated mound of glacial gravel, Knockaunadagrean, has been almost completely dug away. As I mentioned earlier, I am surprised that no cists were reported during the course of its removal as cist-builders 'ought' to have been attracted to it.

Further afield, there is no doubt that there was considerable activity

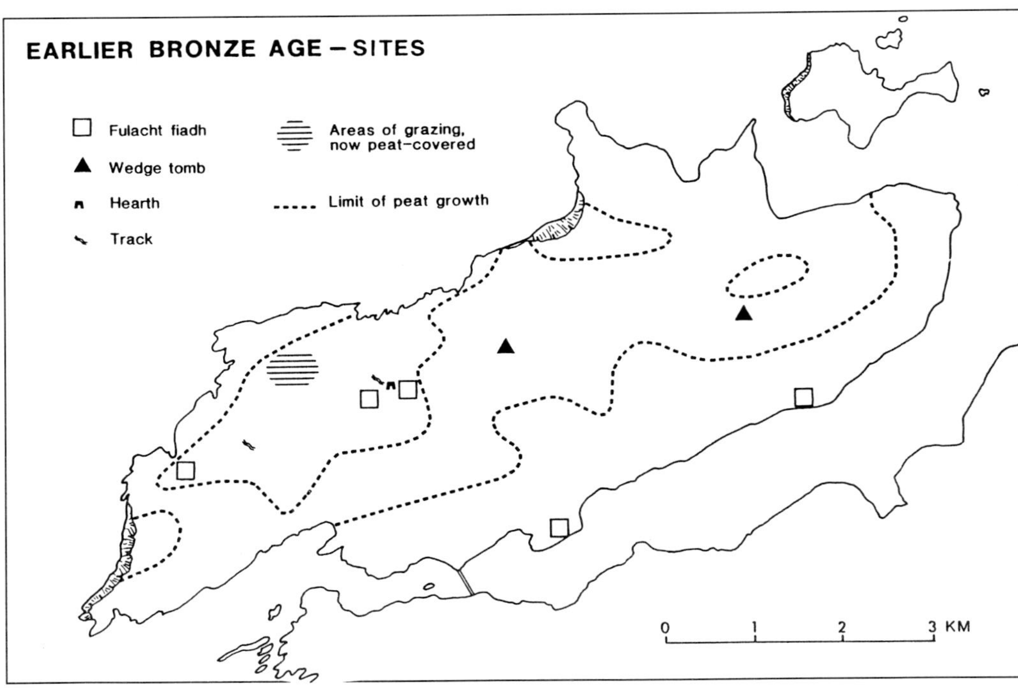

96 *Distribution map of Earlier Bronze Age sites.*

in south-west Ireland during the earlier phases of the Bronze Age. Though there is no evidence for major forest clearance in the immediate vicinity of Cashelkeelty, there are many wedge tombs, generally assigned to the 13
first Beaker phase — opening around 4000 BP — in the surrounding countryside. The South-West Kerry Archaeological Survey has considerably increased the number of such tombs known in that area. Brian O'Kelly dated cooking-places in west Cork to 3700 BP. Extensive copper-mining was carried out near Schull, also in west Cork, and a series of ^{14}C dates obtained there by John Jackson and William O'Brien centres on 3300 BP.

For the Beaker phase the island does have two wedge tombs, a trackway 29
dated to 3930 BP with a stone hearth nearby, and a mass of charcoal 35
and stone probably associated with a *fulacht fiadh*. The charcoal, which 90
was derived from large timbers, had a ^{14}C age of 3800 years. The wedge tombs indicate the presence of a well-organised community.

We have three sites with slightly later dates, all relating to man-made fires. At these sites all the charcoal was derived from twigs and stems of ling (*Calluna*) only; there was no tree charcoal, suggesting that by this time there had already been extensive woodland clearance, and that there must have been further wide stretches of grassland, heath and bog. Two of the sites, dated to 3400 and 3160 BP, were no more than small lenses 67
of charcoal and stone, indicating relatively transient fires. The third, 68
dated to 3250 BP, was a substantial *fulacht fiadh* with stone-lined trough and well-built fireplace. Pollen counts from the site in the Bray West Bog Embayment show evidence of vegetational disturbance at the time 17
of its use.

Moving into the third millennium BP we have three further sites. On the seashore at the bottom of the Laharan Slope were the remains of a disturbed *fulacht fiadh* and its associated charcoal (dated to 2970 BP) in fen-peat; the charcoal was derived from small pieces of wood, not from ling, so there may have been some scrub in the vicinity. The surface layer of peat immediately below an Early Christian *fulacht fiadh* on the Reenarea Rise had a date of 2950 BP, but this does no more than show 83
that by this time there had been extensive formation of peat in the area. Peat at the level of a slab trackway in the Imlagh Basin had a date of 63
2920 BP.

In the earlier phases of the Bronze Age, there must have been extensive areas of closely-grazed grassland on the island. There were still some trees, but woodland was less extensive than it had been when it recovered from the earlier Neolithic attacks. Fen-peat was covering more and more of the lowlands, and blanket-bog was beginning to cap high ground.

LATE BRONZE AGE

I place the opening of the Dowris phase of the Bronze Age, often regarded as the Late Bronze Age proper, at about 2700 BP, when many new influences began to reach Ireland. A bronze sword of this phase is in

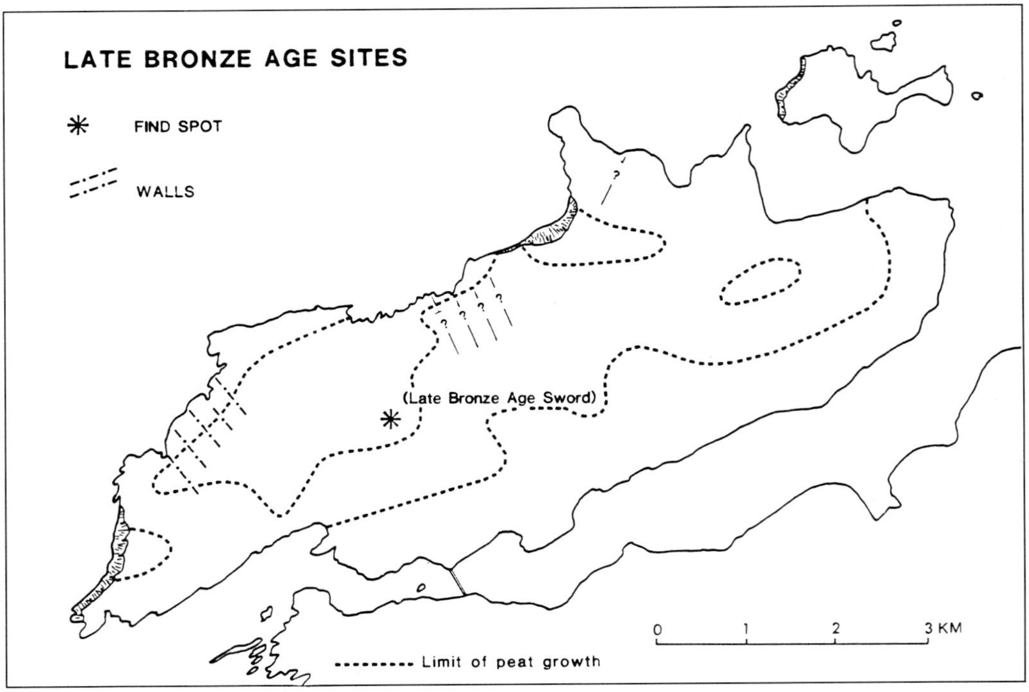

98 *Distribution map of Late Bronze Age sites: areas with walls of less certain age (indicated by a question-mark) are also shown.*

the National Museum (NMI 1893:40). It was found on the surface of the soil below peat in Coarha Beg Townland, and George Eogan has
99 placed it in his sword Class 4, which developed around 700 BC.

52 The wall system on the Coastal Downland is of this phase. The most
55 impressive wall of the series — Wall 13 — had a date of 2680 BP at the point where it ended in the bog in the Imlagh Basin. The walls in
56 the field system nearby (No. 12), which connected with Wall 13, have a comparable age. A stone which was part of the base of a field wall
57 was seated on mor which had a ^{14}C date of 2900 BP, and was buried by peat which had a ^{14}C date of 2050 BP; it is reasonable to expect a considerable time-lag in the overgrowing or swallowing-up of the wall stones by peat growth.

The pollen diagram from the Brendan Rock-knoll, which lies close
15 to the Coastal Downland, shows an impressive rise in disturbance pollen a short distance above the disappearance of the pine at 3000 BP. This rise must be related to Late Bronze Age agricultural activity on the Downland.

Linear walls of large stones also exist in many other parts of the island, and these may also be of Late Bronze Age date. Antoinette Mannion has investigated stone walls below peat on the mainland in the Behy Valley, 30km north-east of Valencia. The base of the peat near a wall was dated to 2430 BP, and these walls may also belong to the Late Bronze Age.

In the Late Bronze Age in Valencia there must have been a considerable social organisation if the attribution of the wall system on the Coastal Downland, and other walls elsewhere, to this period is correct. There must have been enough manpower to collect and arrange huge quantities of large stones in long walls, and enough supervision to see that the walls were laid out in field systems suitable to the type of agriculture practised. Large regular divisions suggest stock rather than tillage. Owing to burning of the ground cover, either by man or by
99 natural agencies, mor continued to develop in the areas not covered by fen-peat and the formation of blanket-bog spread widely. Trees must have been almost entirely eliminated.

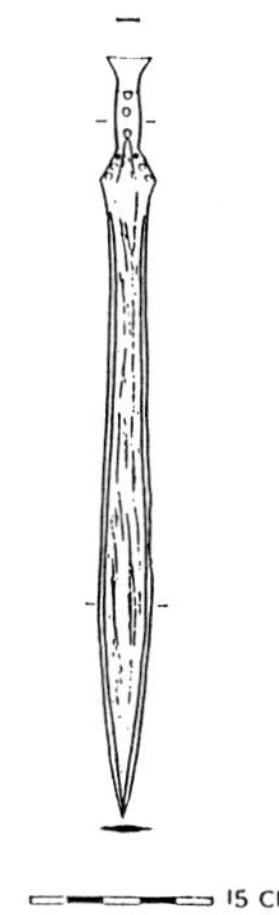

Late Bronze Age sword (NMI 1893:40) from Coarha Beg townland.

IRON AGE

Again our knowledge of this period is deplorably slight. In the Imlagh
52 Basin we have two short lengths of trackways of earth and stone, which may have been constructed to give shortcuts across the bog. In these the largest stone noted was 50cm long, so the scale of construction was much flimsier than that of the walls on the Coastal Downland. Number 18, dated to 1700 BP, immediately beside a large rock-knoll, might lead
71 from the Bray East Slope to the Main Escarpment in the Imlagh Basin, and No.17, younger than 2020 BP, beside a small rock-knoll, might lead

Reenroe, near Ballinskelligs, stones on beach. 99
At certain states of the beach at Reenroe, woody Phragmites-*peat resting on sand is revealed. In places the peat rests on linear arrangements of large stones and clusters of small stones. An arc of large stones is seen here; the overlying peat had a date of 1730 BP, and the stones were probably emplaced in the Iron Age.*

on from the Escarpment towards the sand-knoll, Knockaunadagrean.
75 Near the Brendan Rock-knoll there is a stone trackway (No. 16) buried by almost 1m of peat and dated to 1720 BP.

The radiocarbon dates for these trackways show that there was some activity on Valencia in the Iron Age, though on a more modest level than that of the Late Bronze Age if we base our comparison on the relative sizes of the Iron Age and the Late Bronze Age structures. On the other hand, if my somewhat fanciful suggestion that these trackways are providing shortcuts across the bog is valid, the amount of traffic of the day must have been sufficient to justify the manpower involved in their construction. There must have been a continuing population. Peat, both fen-peat and blanket-bog peat, continued to expand, and the lower ends of the Bronze Age walls were buried by peat.

There is some evidence of other Iron Age activity in the area. On the extensive foreshore at Reenroe, 10km south-east of Valencia, at
99 certain stages of the beach there can be seen lines of large boulders, probably the remains of a field system. These were formerly buried by *Phragmites*-peat, some traces of which still remain. Peat resting on one boulder had a ^{14}C date of 1730 BP. No significant change in sea-level may be involved as on such a shore sand ridges migrate easily. Water impounded by such ridges would be quickly invaded by *Phragmites*, while grass growing on the ridges would give attractive grazing; hence the walls.

Early Christian Valencia

After AD 300 European practices were beginning to become more widely known in Ireland and agriculture appears to have expanded. I open the Early Christian period in Ireland here, and end it at AD 1200, when Anglo-Norman dominance was becoming established. The Barony of *104* Iveragh, in which Valencia is a parish, was defined around 1200. The Fitzgeralds were probably established in Kerry by the year 1250.

The following features can probably be ascribed to this period.

Crosses. Though difficult to date, some crosses are early. The cross on an ogham-inscribed slab in the enclosure at Kildreenagh, which also has remains of rectangular and circular corbelled stone buildings, is obviously in an early ecclesiastical site. The plain crosses at the Brendan Rock-knoll (recently re-erected) became smothered to a depth of 75cm in peat, and could perhaps be dated to around AD 800. The large cross *74* in the Emlagh Bog Embayment — though its status there is doubtful — was hardly imported into the island. Its incised saltire might suggest Viking influence, and Brian O'Kelly has shown that the Vikings were on Beginish around 1050. The cross is very similar in feeling to a large cross at Kilmalkedar, in the Dingle Peninsula. Clochaun A on the Bray South Slope has a Maltese cross incised in the entrance passage, and small crosses and abstract designs are cut on the inner face of its north *45* wall; its excellent drystone corbelled masonry allies it with the typical *46* oratory, usually accepted as early. If its entrance was at the west and not the east end, the building would be assumed to be an oratory.

Ecclesiastical sites. How do we separate an ecclesiastical site from a lay farmstead? Must the former have a cross somewhere? The site marked 'Calluragh' on the Kildreenagh Slope has the ruins of several structures scattered through the innumerable small grave-markers that condemn it to be described as a 'Children's Burial Ground'. No cross has been seen here, but a row of erect slabs are almost certainly the remains of a slab-shrine such as can be seen intact at Killoluaig on the mainland across the Channel. Remains of a building can also be seen in the 'Children's Burial Ground' in Ballymanagh townland on the Laharan Slope. The place-name Kilbeg occurs near here, and I believe there was a church at this site. The same may hold for several other of the 'Children's Burial Grounds', as these sites are named on the Ordnance Survey maps.

I find this blanket-term very unsatisfactory. Although these sites must have been used in the early nineteenth century for the burial of unbaptised infants, I feel that the supply of such children can never have been sufficient to demand the very large numbers of small upright stones that crowd the sites. The cemeteries must earlier have been used for general burial, and the graves were the successors of those of the monks whose monastery originally occupied the site. In some cases traces of the older buildings can still be discerned amongst the hundreds of upright slabs.

Sites with drystone corbelled buildings. I take it that sites with substantial
well-built corbelled buildings are of this period. Such are the well-known
45 clochauns on the Bray South Slope. These lie on some of the best land
on the island, although the cultivation ridges which surround them are
probably later. The corn-kiln which lies further down the Slope may
be of this age, but again may be later. Is it too wild to surmise that these
may be the buildings of a monastic out-farm, on the best land of the
area, which grew food for the neighbouring monasteries? The monastery
2 on the tiny Illaunloghan, 3km east in the Sound opposite Portmagee,
had no land on which to grow food. The land around the monastery
2 on Long Island, lkm to the south-east, shows no sign of having been
35 cultivated. Where did the community on Skellig Michael get its food?
The Cistercians had outlying farms — granges — in the vicinity of their
monasteries; why should the early church in Ireland not have done the
same?

Further similar buildings, but in this case conjoined, lie on the Cool
39 East Lobe, where again the land is good. Here no crosses have been
seen, and the principal building appears to have its door in the north
wall. Is this another outlying monastic farm? A souterrain lies nearby.
The rectangular earth and stone house foundations may be later.

Souterrains. While some souterrains had additional features such as creep-
holes, which would have given the occupants some opportunity for self-
defence, the bulk of them are simple cool storehouses. Every established
farm would have been expected to have one, just as we expect a
refrigerator in every suburban home. There was no need for elaborate
concealment, and several of the Valencia examples appear to have been
built partly above ground level. In two cases the entrance to the souterrain
81 was surrounded by a circular wall, the remains of either a small hut
or a protecting wall. I regard souterrains as of Early Christian age; the
only example I had opportunity to date (on the Cool West Lobe) had
an age of 930 BP, say AD 1000. Most of the souterrains I saw on Valencia
were associated with other structures.

81 *Complex house sites.* Such sites usually have foundations of drystone
42 buildings, one or more souterrains, and remains of a surrounding bank;
43 no crosses are to be seen. I saw three such sites, one on the upper part
of the Oonagranshee Lobe and two on the upper part of the Coarhabeg
Lobe; the Coarha Top rises between the two lobes. The land around
the sites was of good quality and had remained free of peat.

Simple house sites. Valencia showed the ruins of three oval drystone houses (from 8m x 4m to 4m x 3m internally), two of which had a small annexe. The example on Fort Point lay by itself on a rocky slope which showed 108
some terracing; the area was peat-free. The example on the Main Escarpment lay against a low escarpment, and there were two animal 66
shelters and some small rectangular stone-walled fields nearby; the area had been later invaded by peat. The example on the Gortgower Saddle lay near the top of the east-facing slope on well-drained land; it was 6m x 4m internally, and had good-quality walling; it was inside a stone mound about 10m long. Nearby were a ruined cashel, about 30m in diameter, an animal shelter built against a rock outcrop, and the ruins of a circular stone hut, also against rock. Here we must have a small farm of the period.

Raths and cashels. Raths are most common where soils are clayey and of good quality; consequently they are rare on Valencia. On the other hand, the island has slaty rock suitable for building cashel walls. What examples there are of this type of monument have a rather scattered distribution. Two lie on the Gortgower Saddle, about 300m apart. A cashel on the east, which is slightly built, has already been noted as part of a farm complex. A rath, to the west, was more substantial, but has been very much altered, probably during the period of population growth. Much of the bank was demolished, two small rectangular houses were placed inside, and a short length of high massive bank was built to the north, possibly to give some protection from wind. Two are on the Kildreenagh Slope, about 1km apart. The rath to the west, near the end of the Slope, is very much damaged. The cashel to the east is well 32
preserved, shows ruins of interior structures, and has a large standing stone at its entrance. Is this a 'gallaun'? There are five other 'gallauns' on the south slope here within a radius of 1250m. Some 100m to the east are the ruins of two small houses, one of which has a souterrain; there is another isolated souterrain nearby. Have we here the ruins of a village, in which only one house enjoyed the protection of a surrounding bank or cashel? The monument on the Laharan Slope with the inner 88
and outer faces of its encircling bank protected by vertical slabs of stone has already been noted; its age is uncertain. Traces of destroyed circular structures can be seen on the Gortgower Saddle and on the Laharan Slope. The most typical example, a well-preserved rath, is at the south-western seaward corner of the Reenarea Rise, where it almost plays the 83
part of a promontory fort.

Fulachta fiadh. There is a cooking-place on the north-west margin of 84
Reenarea Rise which was excavated by John Sheehan. It had a trough of large slabs and a big spoil-heap, rich in peat-charcoal; it yielded a 85
decorated spindle-whorl and coarse grass-marked pottery (possibly fourth-century AD); the peat on which the spoil-heap lay had a radiocarbon date of 2950 BP.

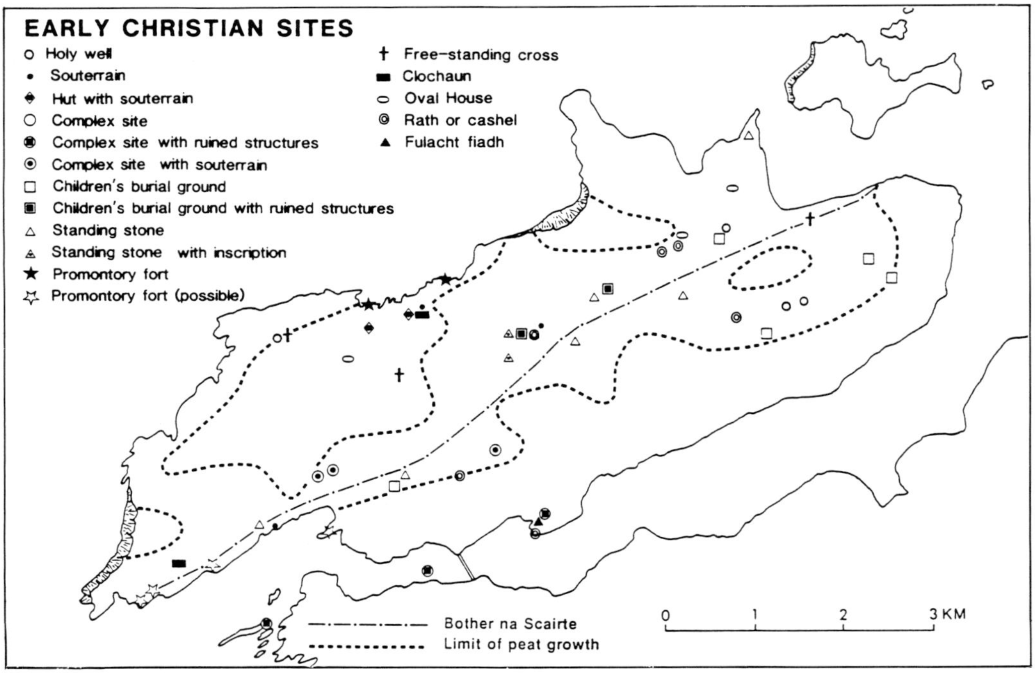

104 *Distribution map of Early Christian sites.*

Gallauns. When is an isolated standing stone a gallaun? Must it have been erected either as an object of veneration or as a territorial marker? When was it erected? Valencia can give no answer to these questions.

In the Kildreenagh ecclesiastical site a standing stone of gallaun size carries an ogham inscription and a later incised cross; is the stone no older than the introduction of ogham? A few hundred metres to the east a stone of gallaun size stands at the entrance to a cashel; is it contemporaneous with the cashel?

There are also typical 'gallauns' standing in isolation. But recently in some cases where heavy walls were being cleared away, the machine operator left an occasional erect stone, too stout even for him to move, standing in isolation, thereby creating a pseudo-gallaun. But if a simple wall was being erected sometime in antiquity, why was it necessary to include such a massive stone in an upright position? Did the line of the wall take advantage of a pre-existing gallaun? At the eastern limit of the Bray South Slope there is an irregular group of gallauns, described by Westropp. Are these the ruins of an impressive stone monument, or are they residual from a former wall?

54 *Promontory forts.* 'Gallauns' appear again in Dundagallaun, on the north-east coast of the Imlagh Basin, where two slender pillar stones stand on either side of the entrance to a small, narrow promontory fort standing out from the cliffs. On the east only a very narrow gully cuts the promontory off from the mainland, and missiles could easily have been

hurled across the gap. Its defensive value must have been low. A short 33
distance to the east, among the cliffs of the Cool East Lobe, we have
Doonroe, a more typical fort. There are several monuments ascribed 54
to the Early Christian period here, and I feel that the fort is also of this
age.

Along the cliffs of the Bray South Slope we have three *dun* names,
Doonroe, Danganbeg and Danganmore Head, although there are no
indications of forts there today. Good land lies upslope from the cliffs,
and I believe that forts, carried away by cliff erosion, formerly existed
here. The now-interrupted cliff-top path, *Bóthar na Scairte,* may have 53
served to connect them. In south-west Kerry heavy cliff-falls are of
frequent occurrence. At Doonbeg in the Dingle Peninsula Terry Barry
has noted that much of the fort had already been carried away by cliff
collapse, and expressed concern that more was likely to follow in the
not too distant future.

This wealth of monuments of the period suggests at first sight that in
Early Christian times Valencia must have had a higher population than
ever before. But we must remember that the continuing expansion of
bog had by now very much reduced the area available for settlement.
Also by this time the rate of bog expansion may have been falling off,
as the higher slopes it was invading became steeper, and so many Early
Christian sites have remained unburied by peat and immediately visible,
whereas the majority of earlier sites were swallowed up. But, as the partly-
buried crosses at the Brendan Rock-knoll demonstrate, some growth of 72
peat did continue even after this period. The Main Escarpment must
have continued to be invaded, and conditions for the occupants of the
oval house there must have become difficult, if not impossible. Bog
covered the coast where Dundagallaun lay, but if the primary purpose
of the fort was defensive the nature of the surrounding land will not have
mattered.

But most of the sites I noted were on well-drained land above the peat limit, and in the Early Christian period on Valencia Island — as throughout south-west Kerry — there must have been substantial populations, both lay and ecclesiastic. It does not seem to have been a period of extensive wall-building.

Historic Valencia

Valencia itself is slow to emerge into coherent history. Church Island, just east of nearby Beginish, produced medieval pottery. In the wall of a shelter, built inside an earlier house, Brian O'Kelly found pieces of a glazed strap-handled jug, which was dated to the late thirteenth or early fourteenth century.

106 Lambeth Library has extensive Carew papers, among them a map of Valencia produced about 1600 after the Desmond rising had been suppressed. The outline of the island is creditable, and of the twelve place-names given all but the two at the extreme west of the island are
47 still in use. One western name, *Crompeol,* could conceivably be the plough-using farm already described on the eastern end of the Bray South Slope, as the name lies in its immediate vicinity. Later in the seventeenth century the name in the form *Crimphole* is attached to the star-shaped fort which was built at the western end of the Dysert Slope in 1653. The name is clearly older than the fort.

Brinamekan is more difficult. The name lies in my West Rock Rim area, and here aerial photographs show clear crop-marks of two small rectangular areas with cultivation ridges. I walked the area several times, but could see nothing on the ground. The name could be *Brigh-na-Meacan,* the hill of some root vegetable or spurge that grew there in profusion. Plant names were used locally for minor features. Thus *Foilatawig* is the cliff of the *Samhadh* or sheep's sorrel; this small dock is a very common weed on acid soils.

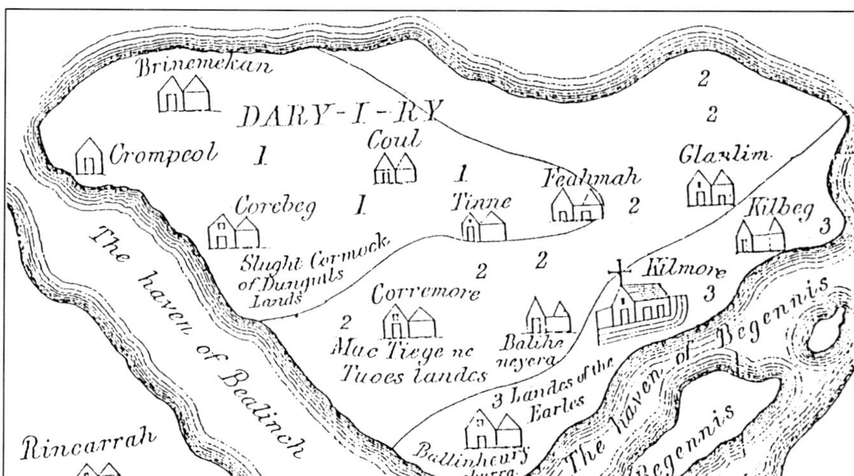

106 *Map of Valencia Island in Carew Papers (c. AD 1600).*

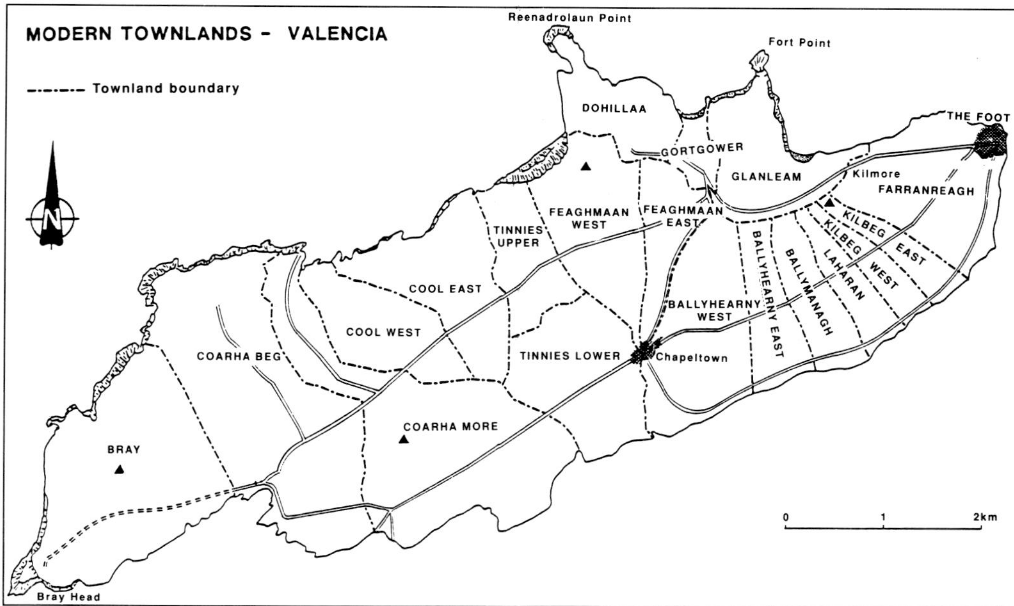

Map of modern townland divisions. 107

The other names are reasonably placed on the map, with the exception of the church, Kilmore, which is too near the south coast. The house-like symbol used for Kilbeg is not surmounted by a cross (as is Kilmore), and the church that gave its name to the locality may have vanished as early as 1600.

The island is divided into three units, which essentially run east-west. 106
Two are still in the hands of branches of the MacCarthy family, who earlier held the whole island, and one is held by the Fitzgerald Knights of Kerry. The Fitzgeralds fared worst, having the still largely bog-covered strip on the south-eastern corner of the island. The western MacCarthys did best, having the good lands of Bray South Slope, the Coastal Downland, the Kildreenagh Slope and the lands around Coarha Top, as well as the valuable turf of the Imlagh Basin. The eastern MacCarthys had the well-drained slopes of the north of the island and the sheltered Glanleam Pocket as well as the poor land of the Chapeltown Basin, the Donnybrook Valley and the Dysert Slope.

In the middle of the seventeenth century star-shaped forts were erected at each end of the Channel to restrict piracy. One was at Crimphole on the south-west near Clynacartan, and the other at the entrance to Valencia Harbour on a promontory now known as Fort Point. 108

The Books of Survey and Distribution, 1636-1703, tell quite a lot about late seventeenth-century changes which altered the east-west divisions into units running north-south, with the exception of the Bray area, which remained a solid block. Coarha Beg, Coarha More and Tinnies cross the island from coast to coast, with Cool on the north wedged in between Coarha More and Tinnies. East of the Chapeltown Valley along the

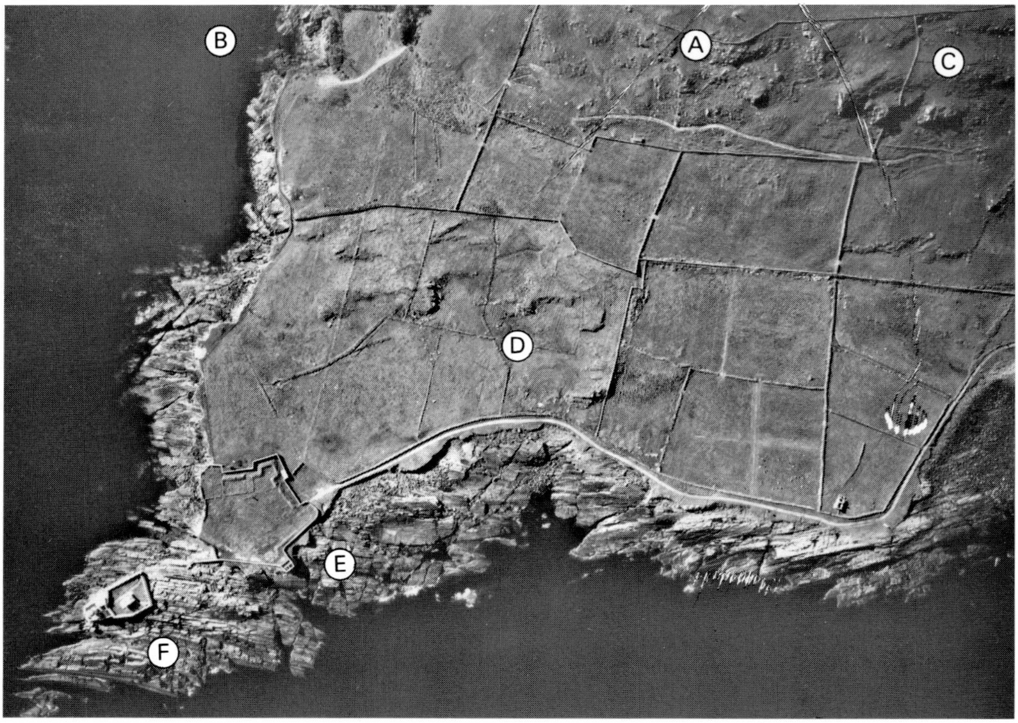

108 *Fort Point.*
On the right at the top we see a relatively high area of improved land (A); to its left, where the level of the ground drops away and there is some shelter, we see the north end of the Glanleam woods (B). The ground then drops away down a terraced rocky slope on which there is an oval house (C), probably Early Christian in age. The land then flattens out in a broad strip of lower ground, and we see a regular field pattern. There is some more rocky ground at the centre of the strip, and here some older enigmatic features survive (D). On the right side of the strip more modern reclamation is clearing away some earlier rectangular field boundaries. At the tip of the Point we see the star-shaped fort (E) erected about 1650 to guard the harbour and the north end of the channel between the island and the mainland. A twin fort was erected at the other end of the channel near Clynacartan. Smuggling and piracy continued until the end of the eighteenth century. On bare rock beyond the fort we see the lighthouse (F) erected about 1840.
Based on Ordnance Survey by permission of the government (permit 5149).

Laharan Slope the picture changes, and the divisions here may be younger. Narrow townland strips run north from the coast over Kilbeg Top to end on the south lip of the Glanleam Pocket. Feaghmaan runs north from the Chapeltown Valley to the high ground of Geokaun Top, and the north tip of the island is divided into Dohilla, Gortgower and Glanleam. At the eastern tip Farranreagh is the equivalent of my Knightstown Nose.

In the early eighteenth century our knowledge of the island becomes very much greater. Trinity College, Dublin, received large grants of land in Kerry in the mid seventeenth century, and at the beginning of the eighteenth was anxious to assess the potential profitability of these lands. A land surveyor from England, Gabriel Stokes, was appointed to survey the lands and assess their value as a source of cash. Stokes was the founder of a family which produced many distinguished *savants,* including the archaeologist Margaret Stokes. Unfortunately his reports make no reference to the potato; it is a pity we cannot know whether this was because at that time it was not much grown in Kerry or whether, not regarding it as a cash crop, Stokes omitted it from mention.

The Trinity lands in Valencia were a broad strip of 770 Irish acres (say 350ha) across the island from north to south in *Culle* and *Tinny,* now the modern townlands of Cool West and Cool East, Tinnies Upper and Tinnies Lower. Stokes reported: 'Coole and Tinny are a very good farm [estate] lying about the middle of the Island Extending from sea to sea The arable yield all sorts of Corn the bogs scarce pasturable but Capable of being Improved by Draining Having a good fall of each side The middle part is mountain and heathy pasture hath on it two good farm houses this farm [estate] is well provided with all necessaries by sea and land haveing plenty of all sorts of sea fish turf firing and fresh water.' One must remember that Stokes probably wanted to please his clients and tended towards rosy descriptions. He had a fine list of adjectives which he would string together in German fashion; a piece of Kerry land, not in Valencia, was described as 'high heathy good dry mountain pasture'.

A map accompanied his report, and on it the word 'arable' only appears on the Kildreenagh Slope, the Cool East, Cool West and Tinnies Lobes (the good land we have already identified), with in addition a small area west of the exit of the Chapeltown river. Here marine erosion has left only the remnants of a low-lying rocky knoll which had well-drained soil; it must have been cleared of peat by 1715. Higher ground is 'mountain pasture', and the low ground of the Chapeltown Valley is 'heathy wet boggy pasture'. The eastern part of the Imlagh Basin that lay within the estate is described as 'red bog'.

Our next information comes forty years later. Smith, in his description of Kerry published in 1756, does not appear to have visited Valencia, but flatters its agricultural potential. He describes it as fertile and an important granary, a statement that has been quoted many times. In my opinion the land is infertile, and corn can never have been widely grown. Indeed, speaking about Cahersiveen on the mainland nearby, Smith says that 'the inhabitants concern themselves very little with tillage, their chief support being potatoes: neither, if they had corn, had they a market near enough to send it'; this was doubly true of Valencia.

Many of Smith's observations about the county are applicable to Valencia. 'The principal products ... are butter, beef, hides and tallow'; 'There is very little corn ... beside what is consumed by its inhabitants, who, some years, are put to great straits for want of bread'. About bogs he tells us that these 'are every year increasing, and rising higher, by the additional rotting and decay of the coarse sedgy grass [presumably

Molinia], and other vegetables that grow on their surface'. He advises that 'these deep stubborn morasses require to be covered with a considerable quantity of coarse gravel, in order to destroy the rank sedgy grass, and render them sufficiently firm for cattle to graze upon: by this method a better and finer type of grass is often produced. When these bogs are 8 or 10 feet deep, as they are in many places, after the drains have been cut and when they subside, and acquire some solidity, paring down the surface with a breast-plough, and burning the sods, so as to destroy the roots of the grass, sedge, moss, and other noxious weeds, is recommended as a proper method to render them profitable for tillage. But in some parts of this country it may be a question, whether the labour and expense, will not be more than the value of the land, after it has thus been reclaimed?' This eighteenth-century question of Smith's is still very relevant today.

Speaking of potatoes he says: 'since the culture of potatoes hath been known ... the herdsmen find out small dry spots to plant a sufficient quantity of those roots for their sustenance, whereby considerable tracts of these mountains are grazed and inhabited, which could not be done, if the herdsmen had only corn to subsist upon'. At the same time he thought that some of the upland enclosures were older, because he says of the southern parts of the Dingle Peninsula that there are 'large tracts of mountain, which have been formerly cultivated up to the top. Several of them, which are but poor barren rocks have great numbers of old inclosures and marks of culture on their sides, which are now neglected; and this is a further circumstance that tends to prove that it hath been better peopled formerly than at present.' He considered that the enclosures had been made to protect cattle from wolves, or to surround 'corn fields, which were more numerous before the importation of potatoes, into Ireland, than at present'.

When Smith was assembling the material for his book on Kerry around 1750, he was standing at an environmental watershed. He tells us that potatoes were now the chief foodstuff, but at the same time looks back to days when corn was still important. The easy availability of the new foodstuff had not yet affected vital statistics, and another thirty-five years had still to run before the population started its disastrous rise. But, as Smith notes, potato patches were already creeping up the hillsides, and we may be sure they were also invading the surfaces of the bogs. In summer widespread patches of a different shade of green would diversify the landscape in a new way.

During my Historic period, Valencia saw dramatic changes. In the course of numerous re-shufflings of landowners the earlier layout of the island in east-west strips was replaced by north-south units. This meant that most townlands took their share of lands of differing potential. The cultivation of the potato was introduced. In some areas reclamation of land and drainage and cutting away of bogs were being actively carried on, but where the bogs were left undisturbed their growth continued. Peat rose still higher round the crosses on the Brendan Rock-knoll.

Population growth

In 1785 the population curve began its dramatic rise and everywhere the call was for more land to cultivate, and the principal object of *114*
cultivation was more potatoes. Valencia still had land to be reclaimed.

More peat could be cleared away from the south-eastern slope of the island. Houses began to appear at the eastern tip of the island, looking across the short distance to Reenard Point on the mainland. Known as *The Nose,* the settlement was later transformed into Knightstown. The Zelva School was built.

Bog surfaces could be invaded and further areas brought into cultivation. We fortunately have three early nineteenth-century records of the process, one of 1800 by O'Brien and two of 1814 by Nimmo and by Radcliff.

Nimmo tells of the equipment available; speaking of the area in general he says that the plough was unknown. All cultivation was done with 'an implement the iron blade of which is about five inches broad, and nine long, with a socket making an obtuse angle with the blade, into which a long wooden handle is fitted: the socket does not cover the base of the handle, which is fitted on by a wedge and nails; the edge admits of the appliance of the right foot only; the handle has no cross-piece to draw it back or turn it. The implement admits not in any way being used as a lever, and may in its action be rather considered a peculiar kind of plough than as a spade. It appears particularly adapted to the cultivation of sloping and stony ground; but its effect in digging or removing the soil being small, it is used in conjunction with a small triangular type of shovel, which is fitted with a handle similar to the spade and employed for spreading the earth, which has been previously stirred by the spade in the trench, over the intervening beds; this being the universal mode of cultivation, at least in this part of Ireland.'

Nimmo is obviously describing the loy, a western type of spade. A version of the loy, the fack (illustrated by Estyn Evans), is still in *112*
occasional use on the island today. He goes on to say that the implement is 'totally unfit for the purpose of cutting turf', and that 'if the cultivator is still to be kept without a plough, cart or team, at least the spade should be as highly improved to him as the instrument will admit'.

Alexander Nimmo was a Scottish engineer who was employed by the Commissioners on the Bogs of Ireland to assist in a general survey of Irish bogs. He produced schemes for the drainage and improvement of bogs, and among many others the bog in the Imlagh Basin drew his attention. He reckoned that there were here 'about 150 acres [105ha] of about 10 feet [3m] deep, wet bog' and noted that 'the edge of this bog over the cliff [the Coastal Downland], seems prevented from

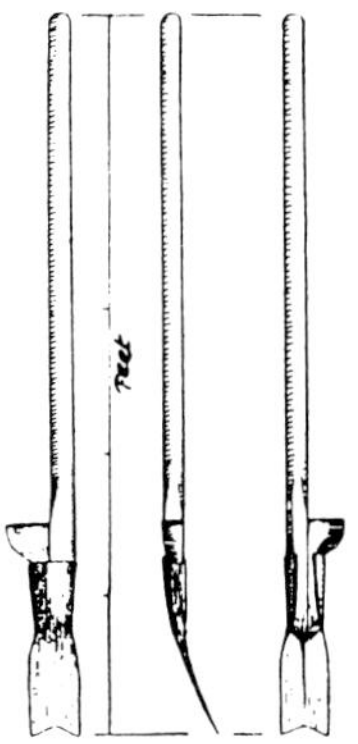
The Kerry 'fack'.

112

increasing by the spray and waves of the Atlantic'. Some of the bog had already been reclaimed and was, in the third year of cultivation, growing a demonstration plot of wheat and flax. Drainage and an access road were needed.

'The Drainage is to be effected by a set of trenches along the upper edge of the bog, to catch the mountain water which now keep it wet, and deliver them into the stream, which, where necessary, are to be scoured and straightened.

'The bog is then to be relieved from surface water by a series of shallow drains, which also deliver into the natural streams, and which are so contrived, as to further, if need be, the business of irrigation. These drains are calculated at two different prices, the one for flat red bogs, containing much water, the other for the firmer black bog, of considerable slope; as I conceive the former will require more numerous cuts than the latter.'

This gives us a valuable picture of the unaltered state of the bog, with ombrogenous peat in the centre and soligenous peat on the surrounding slopes.

The first (1842) edition of the six-inch OS map shows a road stretching out from the lower slope of the Coarha Top as far north as the Main Escarpment, which indicates that Nimmo's recommendations had been followed. At that time the Escarpment must have been entirely buried by ombrogenous peat, as the road across it still rests on 1.80m of firm dry peat, and nearby field banks of uncut peat are at a higher level than the road.

Charles O'Brien is a man of mystery. In 1800 the Royal Dublin Society launched ambitious surveys of the rural economies of all the counties of Ireland. Several county surveys were published, but that of Kerry failed to appear. Henry Pelham was commissioned to carry out the survey, but he died before the work was completed. Nothing is known about O'Brien, but he appears to have cooperated with Pelham, and completed the work. Although the survey follows the RDS pattern, it was not published by the Society, and only appeared in print in 1968.

O'Brien tells us that farming in Kerry centred on grass for summer feed and oats and hay for winter feed for the herds of small black Kerry cows, whose milk provided butter for sale, and on potatoes for human consumption.

Like Stokes, O'Brien has various categories of land; we need only concern ourselves with those types that occur in Valencia. On the uplands he regards wet clayey land as the worst of all, giving poor grass of which one cow requires four to five acres; on the other hand, high bog and heath will carry large numbers of dry stock in summer and autumn. Tall heather and ling are a sign of richer land, which, if given some manure, will produce good crops of potatoes and oats and excellent meadow during the course of rotation; the usual five-year rotation was potatoes: potatoes: oats: oats: grass. For tillage the fields were laid out in ridges from seven to nine feet wide. O'Brien disapproved strongly of the practice of not cutting the meadows before the grass had formed its seeds and its stalk had turned yellow. Encouraged by the Knight of Kerry, some flax was grown; however, the fibre was not woven on the island, but exported as thread to Dingle.

On lower ground there was bog one to two feet thick usually on stiff clay carrying broom, heath and Irish furze (*Ulex europea*); dry, spongy bog (presumably drained ombrogenous peat) which was difficult to bring to a good tilth; and 'fat bog', rich in ashes when burned (this will have been low-lying *Phragmites*-peat). The last could be improved by burning — *burn baiting* — but the improvement would be ephemeral unless followed by the application of sea sand — O'Brien's specific for the improvement of all peats. He says: 'Sea sand has been very little used on bog; where it has its effects have been wonderful. The heath and coarse grass have been destroyed & succeeded by a carpet of the best grasses as Shamrock, ribb'd grass, poa and Timothy.' Failing sand, clay or ditch-scourings would bring substantial improvement.

O'Brien was not too enthusiastic about the use of seaweed — *seawreck*. 'Seawreck may be considered an excellent help to force one good crop, but not an improver of soil, being exhausted in the first season, and where used alone as manure leaving the soil in a poorer state than before, throwing up instead of grass Sheep's sorrell, Moss &c.'

Seaweed is still used in a minor way on the island, and I am sure
it was used together with sand when the tracks down the Laharan Slope *65*
were in use and extensive areas of the bog surface in the Clynacartan Bog Embayment were under cultivation. Thin sheets of sand and fine gravel can still be seen underneath the old lazy-beds on the bog surface here. But there is no evidence today that sand was ever spread in depth, as can be seen in other parts of south-west Ireland. Only thin layers were seen, and there was no trace of surviving shells. Nimmo remarks that such intakes were usually about two acres in extent. Old cultivation ridges can also be seen on the bog surface on the Reenarea Rise and in the Donnybrook Valley.

Thomas Radcliff was a country rector, and his work was sponsored by the Farming Society of Ireland. He tells us that Valencia had 1000 head of cattle and 3000 sheep, and that only small quantities of corn were produced. Tillage was at 'a low ebb, very little being done by the plough, almost all, by manual labour'. The tillage lands had been let out to grass for dairy cows, and potatoes abounded everywhere. On the bogs in Valencia the rotation was six or seven crops of potatoes, one crop of oats and then grass. The bent-grass fiorin (*Agrostis stolonifera*) spreads itself — as its Latin name implies — by a creeping root-stock, and Radcliff recommended that the spontaneous appearance of the grass should be supplemented by scattering further 'fiorin strings' on the ground and covering them with soil. O'Brien and Radcliff agree that bog cultivation was very successful, provided that cattle could be kept out and the drains kept open.

Both Nimmo and O'Brien (and many other observers) speak about

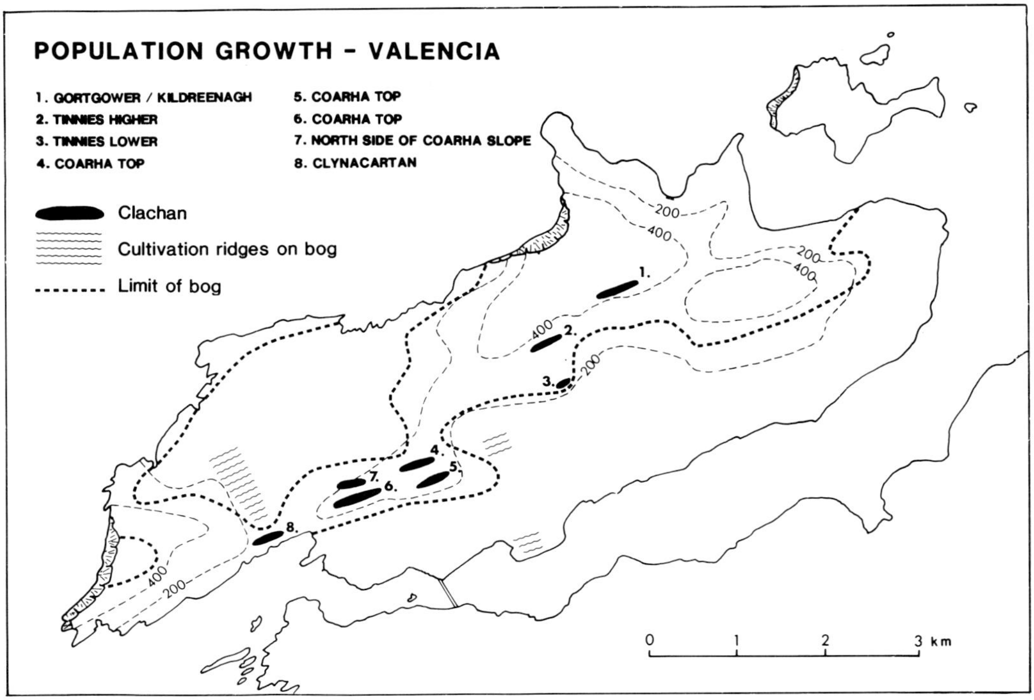

114 *Distribution map of sites of the period of population growth.*

the appalling living conditions of the bulk of the population. In 1845 the Trinity College lands held 147 houses: one of these had at least four rooms, two had two rooms, and the rest had one room only. Diet and clothing were also miserable, but at least the abundance of turf made it possible to keep warm. Some men worked in the slate-quarries, others crewed for the local fishing-fleet, but for most summer emigration as a hired labourer was unavoidable.

At the other end of the scale, Knightstown was developed in the early 1840s, and at that time, in addition to Glanleam House, there were two other gentlemen's 'places' on the island. Donnybrook House lay at the mouth of the Donnybrook River, and Coarhabeg House was opposite Portmagee; both are ruined today.

With an overwhelming demand for ground on which to grow potatoes, where could land for even the meanest of dwellings be found? The answer was on broken rocky ground, often on hilltops. The Coarha Top gives the clearest demonstration of this. It carries three rocky ridges, separated by gullies formerly filled by bog. The south face of each ridge shows a clutter of ruins of small houses and 'gardens', small embanked areas
114 where cabbages and other vegetables were grown. These house clusters had no services of any kind, and are best called 'clachans' in the Estyn Evans sense of the word. The north slope of the Top also had a clachan, and here at Coarhabeg four occupied houses still survive. Nimmo's access road to the Imlagh Basin led down from here, and the inhabitants of

Donnybrook Valley, Chapeltown. 115
This typical chapel-village lies at the junction of the two streams that drain the Valley. The very poor status of the local soils is clearly seen.

the Clynacartan clachan, a little farther west, had easy access to the bog in the Clynacartan Bog Embayment below. At Clynacartan a prettily corbelled well, probably of this age, still survives. Two rock ridges near the top of the Tinnies Lobe also carried clachans. Two clachans had rather better sites on the Kildreenagh Slope, one merging into the upper Tinnies clachan, and a second long straggling clachan at its eastern boundary, which continued on into the Gortgower Saddle.

After Emancipation in the early nineteenth century the Catholic Church began to erect places of worship at strategic sites. It was reckoned that such a church would serve a radius of five miles, then considered to be an easy walking distance. In order to maximise accessibility, crossroads or bridging-points were favoured as sites. *An Caol* (the Narrows), at the point where the rivers draining the Chapeltown Valley met, and little more than four miles from any point on the island, was the obvious choice. The church, once built, immediately became a focus of activity, and there quickly sprang up beside it a school, a public house and a forge, and later a post office. Such 'chapel-villages', as Kevin Whelan has called them, are not uncommon in Ireland, and Caol, now Chapeltown, is a copy-book example of the type. 115

The period of population growth in Valencia was a time of dramatic change. Some larger farms may have remained intact, but much land must have been subdivided to produce more and more small patches. In 1845, just before the Famine, the Trinity College lands were in 155 holdings: 35 of these were of 2-4 acres, 55 were from one acre down. Squalor increased steadily. Bog growth came to an end as the plant cover was drastically damaged. Some bogs were cut away to provide both fuel and new land to cultivate; the surfaces of others were first burnt and then had sand and seaweed spread on them to provide still further land for potato-growing — and food for still more families.

Valencia readjusts and looks ahead

When the Famine struck in 1845, Valencia must have had more than 3000 inhabitants; at the next census in 1851 it still had 2500, a drop of less than 20% or one in five of the inhabitants. Many other parts *118*
of Kerry showed much steeper falls; what was the cushioning effect in Valencia?

Not yet over-fished, the waters around the island offered plentiful harvests. At the beginning of the nineteenth century sufficient peat had gone from The Nose to enable a small settlement to spring up there, opposite Reenard on the mainland, and in 1825 the first pier was built here to accommodate the 500 boats fishing from Valencia. In 1837, according to Lewis's *Topographical dictionary,* 400 people were working in onshore fishery. During the Famine a curing station was set up on the island. At low spring tides — *strands*, as they are locally called — edible molluscs and seaweeds could have been collected. Substantial amounts of food must have been accessible.

After the first post-Famine drop, the rate of population loss fell off, but there has been an uninterrupted — and still continuing — fall, which has brought today's population below the seven hundred level.

In the 1850s the concept of developing the harbour as a deep-water terminal for transatlantic vessels was widely canvassed. The scheme envisaged a railway connection on the mainland at nearby Reenard. But when the railway did finally arrive in 1893, it was too late, and the port never appeared. The railway closed in 1957.

A later transatlantic venture was more successful. The first telegraphic cable came to Valencia in 1865, and as the volume of transmission increased a large number of skilled workers was needed. Handsome terraces of houses and a substantial telegraphic station were built. But *35*
all the workers came from outside the island, and were as isolated from the general community as if they had been living in a British hill-station in India. The fantastic developments in telecommunications that took place during World War II made many cables obsolete, and the station closed in 1965.

As the numbers of the general population fell, pressure for land to grow food fell also and the height to which land on the hills was cultivated reduced in level. Rough grasses and furze invaded the abandoned cultivation ridges, which can still be seen, up to heights of 165m (500ft) in many places. Cultivation of the surfaces of the bogs ended, and the bogs became sources of fuel only. Spades, ploughs, horses and other agricultural accessories gradually came into use and back-breaking labour became less necessary, but the basic pattern of land use was little altered. The reduction in the number of farmers allowed the surviving farms to increase in size, and living standards, including housing, rose also.

One family in 'Paris', as the Bray East Slope is generally referred to on the island, has rehoused itself three times since it was first planted there by the Knight of Kerry in the early 1870s. The Knight provided a slated stone cottage of one room with an attic above, and about 30 acres of land laid out in rectangular fields. The family first erected a small two-storey house in front of the old one, and later a larger annexe was attached to this. This older structure has now been abandoned in favour of a smaller house of bungalow style.

We can perhaps enter a note of caution here. If we are trying to deduce house types from ruins which are several centuries or more in age, we should remember that if a modern family rehouses itself four times in a century, prehistoric families may have done the same. In our studies we often ask ourselves 'What was the duration of the life of a rath?'

Many old ways of life continued on into the middle 1930s. Oared seine-boats still fished for mackerel; only occasional cars and tractors made a perilous trip across the Harbour perched on top of the ferry-boat. But the end of de Valera's 'Economic War' and the approach of World War II brought about a general increase in the rate of economic activity, an increase that reached even to Valencia. The outboard engine replaced the oar.

In the last thirty years a dramatic change in land use has taken place world-wide, a change consequent on Harry Ferguson's invention and development of machinery powered by hydraulic pressure. The tractor was no longer an 'iron horse', but it, and the larger machines which flowed from it, could perform a wide variety of operations, not least earth-moving in bulk. Its island status had kept large machines from Valencia, but this impediment was removed when the bridge to Portmagee was opened in 1970. Today Valencia has no working horses, but it has got a depot for the maintenance of earth-moving monsters.

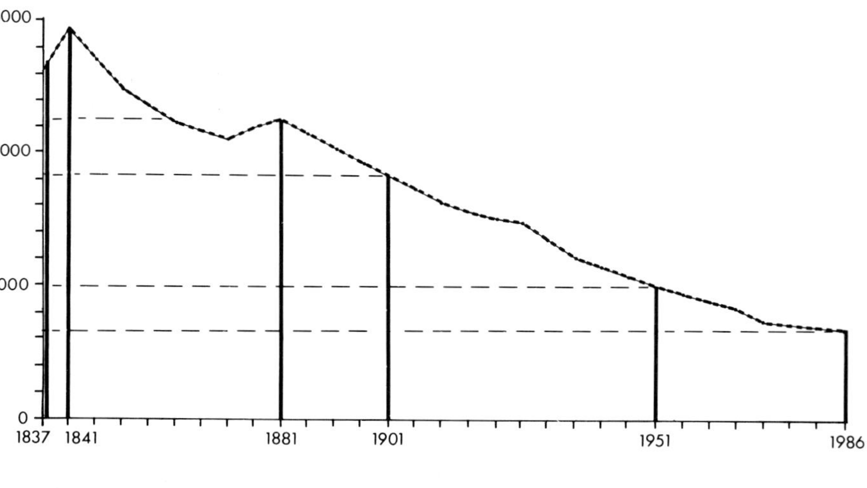

118 *Population curve for Valencia Island, 1837-1986.*

After Ireland joined the European Economic Community in 1973 the prices paid to Irish farmers for their commodities rose sharply, and when the Community began to make generous grants for land reclamation, enterprising farmers on the island were well placed to exploit the possibilities. Rougher and higher tracts of land that had not been cultivated for over a century were smoothed and drained, fertilised, and laid down to grass to produce milk. Considering the richness in antiquities of those parts of the island that have escaped reclamation, it is clear that many sites must have been lost in the process of clearing. Grassland for milk has become the order of the day, and tillage has been completely abandoned.

The bridge brought about other dramatic changes. Knightstown effectively died, without bringing any compensating revival to Portmagee, which is skirted by the approach road to the bridge. The present hostel status of the Royal Hotel in Knightstown compared with its glories of one hundred years ago, when it received princes and poets, epitomises the position. Local legend has it that Tennyson wrote 'Crossing the bar' when staying in the hotel. Motor-cars no longer brave the perils of the Harbour; instead they daily sweep in their dozens to the supermarkets of Cahersiveen. For a population of some six hundred and fifty (men, women and children) the island has about two hundred active cars, as well as the many deceased ones which repose in various parts of the island.

Today the islanders live in rural surroundings with most beautiful scenery on all sides, but the younger generations do not live a rural life in the old-fashioned sense of the word. In 1800 O'Brien wrote that 'Hogs are fed here in great numbers, every cabbin even the poorest rearing at least one'; today I think there is not a single pig on the island. Farmyard hens have virtually disappeared, except where there is a grandmother to tend them. Television gives ample contact with the outside world. Island life in one sense is almost suburban rather than rural, with the focus on Cahersiveen and towns further afield. The school bus carries the island children to secondary school in Cahersiveen. But if insularity has been technically abolished remoteness has not, and the island has little to offer to its young people in the way of job opportunity.

What can the establishment of the Single European Market do to rectify this state of affairs? It is not easy to see when or how the population drain can be checked. New houses are being built, some by parents who hope to retain their children on the island, others by speculators for rent as holiday homes. These ventures may or may not succeed. Irish holiday-makers want easy access to sunny beaches, and there is no first-class beach on the island. Sea-fishing and scuba-diving are good. The clear water around the island offers excellent opportunities for underwater photography. Emptiness and opportunities for walking have their own attraction, not for many Irish people but for many from industrialised, overcrowded western Europe. These people already come to visit, and some come to settle; more will probably come. Already several houses are owned by Germans, both permanent residents and occasional visitors;

the former home of the Knight of Kerry is in such hands. A judicious encouragement of a certain specific type of tourist trade might reap a fair reward.

But while population continues to fall, pressure on the environment is slowly being reduced. The EC is painfully realising that it has too much land, and there will be no further grants for land reclamation. General environmental interest is slowly growing in Ireland, though it has a long way to go before it reaches the general level of the Community. Valencia is fostering that growth, as it has an active Heritage Centre in Knightstown. When the listing of antiquities at present being conducted by the South-West Kerry Archaeological Survey is complete, then there will be a greater degree of protection for the precious record of former environments that Valencia can still provide, more richly than many other parts of Ireland.

Bibliography

ANON. 1846 *The Parliamentary Gazetteer of Ireland.* London.

BARNOSKY, Cathy W. 1988 A late-glacial and post-glacial pollen record from the Dingle Peninsula, County Kerry. *Proceedings of the Royal Irish Academy* **88** B, 2-37.

BRYANT, R.H. 1974 A Late-Midlandian section at Finglas River, near Waterville, Kerry. *Proceedings of the Royal Irish Academy* **74** B, 161-78.

CLEARY, Rose 1985 A Bronze Age cist at Pound, Portmagee. *Journal of the Kerry Archaeological and Historical Society* **18,** 215-20.

COXON, Peter 1985 Botanical appendix. In Mitchell 1985, 35-47.

DELAP, M.J. 1911 Holy wells: Valencia and Portmagee. *Kerry Archaeological Magazine* **7,** 403-14.

DE VALERA, R. and Ó NUALLÁIN, Seán 1982 *Survey of the megalithic tombs of Ireland.* Dublin. Stationery Office.

EOGAN, George 1965 *Catalogue of Irish bronze swords.* Dublin. Stationery Office.

EVANS, E. Estyn 1967 *Irish heritage.* Dundalk. Tempest.

GUILCHER, André 1966 Les grandes falaises et mega-falaises des côtes sud-ouest et ouest de l'Irlande. *Annales de Géographie* **75,** 26-38.

HENRY, Françoise 1957 Early monasteries, beehive huts, and dry-stone houses in the neighbourhood of Caherciveen and Waterville (Co. Kerry). *Proceedings of the Royal Irish Academy* **58** C, 45-160.

HERITY, M. 1970 The prehistoric peoples of Kerry: a programme of investigation. *Journal of the Kerry Archaeological and Historical Society* **3,** 5-14.

JACKSON, John S. 1984 The age of primitive copper mines on Mt Gabriel, west County Cork. *Journal of Irish Archaeology* **2,** 41-50.

JESSEN, K. 1949 Studies in late Quaternary deposits and flora history of Ireland. *Proceedings of the Royal Irish Academy* **52** B, 85-290.

LAMBETH LIBRARY Various Carew Manuscripts.

LAVELLE, Des 1977 *Skellig; island outpost of Europe* (2nd edn). Dublin. O'Brien Press.

LEWIS, S. 1837 *A topographical dictionary of Ireland.* London. Lewis.

LYNCH, Ann 1981 *Man and environment in S. W. Ireland.* British Archaeological Reports, British Series **85.** Oxford.

MACARTNEY ROBBINS, J. 1942 Brief notes on monuments on Valencia Island. Unpublished letter in National Museum of Ireland.

McCARTHY, J.P. 1986 Unpublished fieldnotes on souterrains examined in Valentia Island, April 4-7th, 1986.

MANNION, Antoinette 1983 A note on buried prehistoric field systems near Killorglin, Co. Kerry. *Irish Naturalists' Journal* **21**, 93.

MITCHELL, Frank 1985 *Landscape archaeology in south-west Valencia Island: a field guide* (2nd impression). Privately published.

MITCHELL, Frank 1986a *Landscape archaeology in south-west Valencia Island: a field guide: part 2*. Privately published.

MITCHELL, Frank 1986b *The Shell Guide to reading the Irish landscape*. Dublin. Country House.

MITCHELL, G.F., COXON, P. and PRICE, A. 1983 *Guide book to north-west Iveragh, Kerry*. Dublin. Irish Association for Quaternary Studies.

MOLLOY, Karen and O'CONNELL, Michael 1988 Neolithic agriculture — fresh evidence from Cleggan, Connemara. *Archaeology Ireland* **2** (2), 67-70.

MOYLES, M.G. and DE BRUN, P. 1968-9 Charles O'Brien's agricultural survey of Kerry, 1800. *Journal of the Kerry Archaeological and Historical Society* **1** (1968), 73-100; **2** (1969), 108-32.

NIMMO, A. 1814 Report on the bogs in the Barony of Iveragh. *Commissioners on the Bogs of Ireland, Report 4, Appendix 5.*

O'BRIEN, C. A view of the state of agriculture in the County of Kerry. *Anno* 1800. See MOYLES and DE BRUN 1968-9.

O'BRIEN, William F. 1987 The dating of the Mt Gabriel-type copper mines of west Cork. *Journal of the Cork Historical and Archaeological Society* **92**, 50-70.

O'CONNELL, Michael (in press) The origins of lowland blanket-bog.

O'KELLY, M.J. 1956 An island settlement at Beginish, Co. Kerry. *Proceedings of the Royal Irish Academy* **57** C, 159-94.

O'KELLY, M.J. 1958 Church Island near Valencia, Co. Kerry. *Proceedings of the Royal Irish Academy* **59** C, 57-136.

PEARSON, G. *et al.* 1986 High Precision ^{14}C measurement etc. *Radiocarbon* **28**, No. 2B.

POCHIN MOULD, Daphne 1978 *Valentia: portrait of an island*. Dublin. Blackwater.

SHEEHAN, John 1988 *Preliminary report on monuments of Valencia Island*. Tralee. AnCO/South-West Kerry Development Organisation.

SMITH, A.G. and CLOUTMAN, E.W. 1988 Reconstruction of Holocene vegetation history in three dimensions at Waun-Fignen-Felen, an upland site in South Wales. *Phil Trans RSL,* B, **322**, 159-219.

SMITH, C. 1756 *The antient and present state of the County of Kerry*. Dublin. Smith.

TRINITY COLLEGE, DUBLIN, LIBRARY Various Valencia Island manuscripts.

WESTROPP, T.J. 1912 Notes on the promontory forts and similar structures of Co. Kerry. Part V. Iveragh (Valencia to St Finan's Bay). *Journal of the Royal Society of Antiquaries of Ireland* **42**, 285-324.

WHELAN, Kevin 1983 The Catholic parish, the Catholic chapel and village development in Ireland. *Irish Geography* **16**, 1-15.

WILLIAMS, Brian 1983 Early Christian landscapes in County Antrim. In T. Reeves-Smyth and F. Hamond (eds), *Landscape archaeology in Ireland,* 233-46. British Archaeological Reports, British Series **116.** Oxford.

WOODMAN, Peter C., DUGGAN, M.A. and McCARTHY, A. 1985 Excavations at Ferriter's Cove: preliminary report. *Journal of the Kerry Archaeological and Historical Society* **17,** 5-9.

Glossary

Bronze Age, Beaker A phase with thin-walled decorated pottery, objects of copper and gold, and wedge-shaped tombs, beginning about 2000 BC

Dowris A phase with abundance of metal, both gold and bronze, fabricated in new styles from Britain and Scandinavia, beginning about 700 BC

Pygmy Cup A small decorated vessel associated with Urns

Urn A phase with large coarse decorated pottery associated with cremation and many types of bronze objects, but little gold, beginning about 1700 BC

Calibration The reconciliation of radiocarbon years with calendar years

Clachan A nineteenth-century cluster of simple dwellings without any services, such as church, school or shop

Cleavage A tendency to split easily in one direction, developed in a rock as a consequence of earth pressures

Clochaun A small rectangular building of Early Christian date, erected without the use of mortar

Conglomerate A rock which consists of an aggregate of pebbles in a matrix of finer material

Corrie An armchair-shaped hollow excavated by glacial erosion into a mountainside

Dip-slope A long gentle slope, following in general the dip or tilt of the layers of sedimentary rock underlying it

Etch-plain A rock surface, usually relatively flat, revealed when an overlying mantle of deeply weathered rock has been removed by erosion

Freeze-thaw action In climatic regimes where temperature repeatedly rises and falls through freezing-point, frost expansion splits rocks, and thawing allows loose material to move downslope

Gley Gley forms where a soil is very rich in small clay particles; water cannot drain down through it, and the ground is waterlogged for long periods

Hogback ridge A ridge composed of a series of low rounded rises

Horst A block of rock uplifted by earth pressures, and usually bounded by faults

Inselberg An isolated rock-knoll which stands up above an etch-plain

Old Red Sandstone A rock formation of shales, sandstones and conglomerates formed by the lithification of torrential sands and gravels deposited many millions of years ago

Podzol In coarse-grained non-calcareous soils descending water dissolves iron from the surface layers and deposits it lower down; the surface layers become ash-grey in colour, and the Russian word for ash — zola — is incorporated in the soil name

Pool and hummock The alternating development of pools rich in *Sphagnum* and hummocks rich in ling gradually raises the bog surface

Quaternary The last great period into which geological time is divided; it began about two million years ago, and is characterised by alternation of climate between cold and warm

Scarp-slope A short relatively steep slope where the ground surface cuts across sloping beds of rock; an escarpment is commonly formed along the line where a scarp-slope intersects a dip-slope

Slate A fine-grained rock which splits readily in one direction as a consequence of having been subjected to earth pressures

Snow-bed A body of winter snow which remains unmelted in summer, owing to the aspect of the ground on which it lies

Solifluction The process of unconsolidated soil layers moving downslope under the influence of gravity; the process is greatly accelerated by freeze-thaw action

Tectonic Structures produced by earth movement

Tertiary The penultimate great period of geological time, occurring between seventy million and two million years ago; the climate was warm, and deep weathering of rock took place

Index

PHYSIOGRAPHIC UNITS

Numerical order

1	Dohilla Slope
2	Fort Point
3	Glanleam
4	Kilbeg Top
5	Gortgower Saddle
6	Geokaun Top
7	Fogher Mega-Cliffs
8	Kildreenagh Dip-Slope
9	Tinnies Scarp-Slope
10	Cool Scarp-Slope
11	Cool East Lobe
12	Cool West Lobe
13	Coarhmore Saddle
14	Coarha Top
15	Coarbabeg Lobe
16	Clynacartan Saddle
17	Bray East Slope
18	Bray South Dip-Slope
19	Bray Top
20	Bray Mega-Cliff
21	Bray North Scarp-Slope
22	West Rock Rim
23	Coastal Downland
24	North Rock Rim
25	Imlagh Basin
a	Main Escarpment
b	North Rock Rise
c	Brendan Rock-Knoll
d	Emlagh Bog Embayment
e	Clynacartan Bog Embayment
f	Bray West Bog Embayment
g	Cool Bog Embayment
26	Dysert Slope
27	Oonagranshee Lobe
28	Reenarea Rise
29	Donnybrook Valley
30	Tinnies Lobe
31	Chapeltown Valley
32	Ballyhearny Lobe
33	Laharan Dip-Slope
34	Knightstown Nose
35	Kilmore Shelf

Alphabetical order

Ballyhearny	32
Bray East Slope	17
Mega-Cliffs	20
North Scarp-Slope	21
South Dip-Slope	18
Top	19
West Bog Embayment	25f
Brendan Rock-Knoll	25c
Chapeltown Valley	31
Clynacartan Bog Embayment	25e
Saddle	16
Coastal Downland	23
Coarha Top	14
Coarhabeg Lobe	15
Coarhamore Saddle	13
Cool Bog Embayment	25g
Cool East Lobe	11
West Lobe	12
Scarp-Slope	10
Dohilla Slope	1
Donnybrook Valley	29
Dysert Dip-Slope	26
Emlagh Bog Embayment	25d
Fogher Mega-Cliffs	7
Fort Point	2
Geokaun Top	6
Glanleam Pocket	3
Gortgower Saddle	5
Imlagh Basin	25
Kilbeg Top	4
Kildreenagh Scarp-Slope	8
Kilmore Shelf	35
Knightstown Nose	34
Laharan Dip-Slope	33
Main Escarpment	25a
North Rock Rim	24
Rise	25b
Oonagranshee Lobe	27
Reenarea Rise	28
Tinnies Lobe	30
Scarp-Slope	9
West Rock Rim	22